Utility Marketing Strategies:

Competition and the Economy

Utility Marketing Strategies:

Competition and the Economy

By: Clark W. Gellings

Published by
The Fairmont Press
700 Indian Trail
Lilburn, GA 30247

Library of Congress Cataloging-in-Publication Data

Gellings, Clark W.
  Utility marketing strategies: competition and the economy / by Clark W.
Gellings.
    p. cm.
  Includes bibliographical references and index.
  ISBN 0-88173-156-0
  1. Electric utilities--United States. 2. Electric power--United States--
Marketing. I. Title.
HD9685.U5G454   1994   333.79'323--dc20   93-25333
                                                       CIP

Utility Marketing Strategies: Competition and the Economy
by Clark W. Gellings.
©1994 by The Fairmont Press, Inc. All rights reserved. No part of this publication may be reproduced or transmitted in any form or by any means, electronic or mechanical, including photocopy, recording, or any information storage and retrieval system, without permission in writing from the publisher.

Published by The Fairmont Press, Inc.
700 Indian Trail
Lilburn, GA 30247

Printed in the United States of America

10 9 8 7 6 5 4 3 2 1

ISBN 0-88173-156-0   FP

ISBN 0-13-927757-9   PH

While every effort is made to provide dependable information, the publisher, authors, and editors cannot be held responsible for any errors or omissions.

Distributed by PTR Prentice Hall
Prentice-Hall, Inc.
A Paramount Communications Company
Englewood Cliffs, NJ 07632

Prentice-Hall International (UK) Limited, London
Prentice-Hall of Australia Pty. Limited, Sydney
Prentice-Hall Canada Inc., Toronto
Prentice-Hall Hispanoamericana, S.A., Mexico
Prentice-Hall of India Private Limited, New Delhi
Prentice-Hall of Japan, Inc., Tokyo
Simon & Schuster Asia Pte. Ltd., Singapore
Editora Prentice-Hall do Brasil, Ltda., Rio de Janeiro

TABLE OF CONTENTS

Dedication ix
Preface xi

SECTION I. INTRODUCTION: ELECTRICITY'S VALUE AND THE ATTRIBUTES THAT CREATE THIS VALUE

Chapter 1. Electricity's Value:
A Current Assessment 3
   Why Is Electricity So Valuable? 3
   A Historical Perspective: Utility Marketing Efforts 7
   The Environment, Politics, and Competition
      Affect Perceptions of Value..9
   What Can Utilities Do to Enhance
      Electricity's Value to Customers?....................................11
   Some Current Examples of Value-Based
      Programs/Services..13
   A Few Overall Suggestions for Improving
      Value-Based Marketing Efforts......................................15

Chapter 2. Electricity's Attributes 19
   Introduction 19
   Attributes of Electricity 20
   Total Resource Efficiency 26
   New Opportunities for Electrotechnologies 33

SECTION II. ELECTRICITY'S ENVIRONMENTAL, POLITICAL, SOCIAL, AND ECONOMIC IMPACTS

Chapter 3. Environmental Impacts and
New High Value Solutions 55
   Environmental Impacts of Energy Production 55
   Existing Electrically Powered Technologies Meet Stiff
      Environmental Challenges 60

Chapter 4. Electricity's Relationship to
U.S. National Security...79

The Global Marketplace
Deemphasizes National Borders 80
The Environment as a Tool and Weapon 82
U.S. Energy Dependency: Oil Imports................................83

Chapter 5. Electricity's Impact on Family Life **93**
Electricity's Introduction into Domestic Life 94
Social Consequences of the
Introduction of Electricity 98

**Chapter 6. The Links Between Electricity,
Energy Efficiency and Economics** **107**
Energy Efficiency and Its Link with Electricity 109
The Economics of Energy Efficiency 114
A Look at Future Energy Use
and Efficiency Issues...120

Chapter 7. Technology's Impact on Energy Efficiency **129**
The Importance of New Technology Developments 129
Factors Affecting Industrial Productivity..........................136
Trends in Industrial Productivity:
A Situation Analysis 141
The Importance of New Technology and
Productivity Increases: A Lesson from
the Steel Industry...142
The Electric Utility as a Technology/Productivity
Bridge for Industrial Consumers 144

SECTION III. ELECTRICITY AND INDUSTRY

Chapter 8. Introduction: Electricity and Industry................**149**
Industrial Electricity Use: A Historical Perspective..........149
Overview of Recent Industrial Electricity Demand 155
Electricity's Future Role in the
Industrial Energy Market..164
Changes in Marketing to Industry......................................166
Summary...171
Chapter 9. Case Studies in Four Industries...........................**173**
Electrification in Pulp and Papermaking 173
Glassmaking..181
Petroleum Refining..186

Agriculture 193
Chapter 10. Electric Transportation 207
   Reasons for Growing Transportation Crisis 207
   Electric Vehicles: One Solution 210
   Transportation Tomorrow 214

SECTION IV. STRATEGIES FOR THE FUTURE

Chapter 11. Customers' Needs 221
   Customer Needs: Basic and Derived 221
   Utility Customer Expectations:
      Findings from a Recent Study 223
   Factors Affecting Customer Behavior 225
   Residential Customer Preference and Behavior:
      What Motivates Residential Customers to
      Make Energy Purchases? 229
   Commercial Customer Preference and Behavior:
      New Research Findings About Commercial
      Customers' Energy Needs 237
   Case Studies: Products/Services Devised to
      Meet Customers' Needs 250
**Chapter 12. Service Marketing in the Electric
Utility Industry: Next Steps** 257
   Introduction 257
   Devising Service Strategies 258
   Understanding Customers' Needs 260
   The Introduction and Evolution of DSM Planning 263
   An Additional Planning Challenge:
      Regulated Marketing 268
   A Few Examples of Services with Added Value 271
   Conclusions 274

Index 277

DEDICATION

Our Love of Amber

Humans have known since ancient times that the organic mineral amber possesses unique qualities. The natural scientists and philosophers of ancient Greece were particularly interested in the fact that when rubbed, a piece of amber would attract light bodies—it became notably electric. The Greeks named this resinous mineral "elektron," from which derives the English word "electricity."

Amber, also called succinite (from the Latin Succinum), is actually the fossilized resin of extinct cone-bearing trees that lived thousands of years ago. It is found on beaches primarily adjacent to the Baltic Sea. Amber is usually yellow in color, but can be found in hues of red, brown, or clouded white. It is resinous in luster, always translucent and sometimes transparent, brittle, easily cut, fused, and burns with a yellow flame. Amber is also highly electrical. The early Greeks discovered they could generate a tremendous amount of static charge when they rubbed pieces of this mineral with cloth or fur.

The patterns of our daily lives depend on electricity. It powers basic equipment as well as the conveniences that help us function in today's society. It is no wonder we are intellectually drawn to electricity—it is the basis of our very bodies. Electric impulses from the brain signal the movement of muscles in our bodies, direct signals of pain, pressures—even pleasure. Electricity shocks us, sustains us, and controls our every human action.

Why is electricity so infinitely superior to chemical energy forms? Electricity embodies the basis of all energy—the movement of an atomic particle, an electron. Electricity can be directly converted, infinitely controlled, and utilized more efficiently and effectively than any other energy form. Fossil fuels, on the other hand, are primitive energy forms comprised of chains of carbon and hydrogen atoms. These chemical compounds must be oxidized to produce heat and yield useful energy—a terribly inefficient conversion that releases harmful pollutants. Once converted into heat, fossil fuels often undergo distribution losses or additional transformation before ultimate utilization.

No other energy form has contributed so much to our very existence. In today's society, we take for granted the availability of the conveniences electricity provides—transportation, illumination, space conditioning, communications, motor power, process heating, computers, and medical diagnostics.

In brief, man would not have gone to the moon, developed computers, automobiles, airplanes, released women into society, or even reared children without Amber—we truly love her.

PREFACE

In 1973, Charles Ross stated "...Ever since the Industrial Revolution, the world, and particularly the United States, has worshipped at the altar of the machine and its indispensable handmaiden, electricity."[P-1] In little over 100 years, electricity has drastically altered the lifestyles of all Americans and most of the world.

Electricity is not a fuel—it is a uniquely refined energy form. Electricity heats, cools, and lights our homes and businesses, refrigerates our food, runs electric motors, facilitates the use of advanced medical diagnostic tools, and powers mass communication systems. Electric processes capture energy from the sun, wind, water, steam as well as the interaction of atoms. When properly applied, electricity provides the most basic services as well as the most technologically advanced with greater efficiency and effectiveness than any energy alternative.

Between 1950 and 1972, electric power consumption surged more than 350 percent; concurrently, the population grew by only 37 percent. During this same period, natural gas and crude petroleum use rose much less rapidly, only 265 percent and 110 percent, respectively. Today, electricity accounts for 36 percent of total U.S. energy use and 10 percent of global energy use. By 2000, researchers estimate this share will grow to 42 percent of U.S. energy use. The global percentage will continue to grow as developing nations become more productive, adopt new technologies, and incorporate modern conveniences into their changing lifestyles. However, electricity's growth is not automatic—electric utilities face fundamental conflicts as they serve electricity consumers in today's energy marketplace.

These conflicts mostly stem from consumers' growing demands for energy using appliances and equipment. As utilities seek to meet these needs, they face twin problems: electricity production, delivery, and use heavily influences the environment and electricity generation requires substantial capital investment in plants and processes. Both problems have generated a great deal of publicity—a lot of it negative—about the environmental, regulatory, economic, and political issues involved. This negative publicity has made many consumers (i.e., electric utility customers) skeptical of their local utilities and the services they provide.

As a result, many of today's electricity consumers do not hold a particularly high perception of electricity and its overall "value" to their lives, when compared with other goods and services. This perception is unfortunate, since electricity either provides or powers most of these goods and services. In addition, electricity is the only energy source that can efficiently reduce overall energy use while still supplying all of a consumer's needs.

Utilities are beginning to recognize that perception—not reality—drives most consumer actions. If consumers perceive they get something valuable for their hard-earned dollars, they are more likely to spend those dollars. Utilities must work to counter the negative perceptions that exist about electricity and electricity production. This process includes resolving many of the current environmental, regulatory, and political conflicts utilities face. Many nonutility businesses have reaped the benefits of creating positive perceptions: if they increase consumers' perceptions of value, they build sales and improve customer satisfaction. For example, Federal Express sells a product (reliable, overnight mail service) that businesses now depend on—and are willing to pay for—despite its high price tag.

Utilities must become value-oriented, instead of product-oriented. Value-oriented companies convince customers that their needs come first. For electric utilities, this transition involves a "giant" leap, a move from old business strategies (focusing on planning and operations) to new ones (developing energy service businesses that meet customers' needs).

Utilities must define and illustrate electricity's value to consumers in terms consumers understand. Yes, addressing a concept such as "customer perceptions of electricity's value" is inexact, qualitative, and exceedingly difficult. However, for the electric utility industry to grow and prosper, it must begin to remind its customers of electricity's extraordinary benefits in simple, direct ways. If the industry does not begin this process, consumers will continue to be reminded of this lesson in a highly negative way—when they flip the switch and discover the power is out. This book examines these different "value"-related issues in greater detail, including electricity's:

- Attributes
- Impacts (on history, the environment, politics, family life and economics)
- Use by world industry
- Role in meeting electric utility customers' needs

Finally, this book offers some practical "how to" advice to the electric utility industry. It discusses how electric utilities can become more value-oriented, increase the value of electricity to their customers, incorporate this perspective into planning processes, and enhance their marketing efforts.

References:

P-1 C.R. Ross, "Electricity as a Social Force," *The Future Society: Aspects of America in the Year 2000*, The Annals of the American Academy (of Political Science), Stanford University Press, 1973.

SECTION I

INTRODUCTION:

Electricity's Value and the Attributes That Create This Value

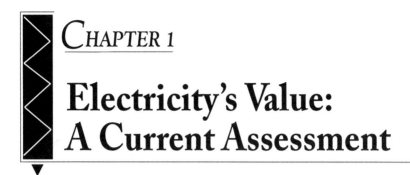

Chapter 1

Electricity's Value: A Current Assessment

Why Is Electricity So Valuable?

Why is electricity valuable? Because it provides services? Facilitates product development and use? Engenders technology? Of course, but its value is not limited to narrow examples of its use. When attempting to quantify electricity's value, it helps to look at the "big picture" for a moment. From this vantage point, quantifying electricity's value may seem less daunting.

While the average utility customer instinctively understands the value of flipping a switch and receiving instantaneous light, this customer may not recognize additional evidence of its value. For example, does the average utility customer (or utility staff member, for that matter) appreciate electricity's impacts on U.S. geographic, technological, and economic development? Early central generating station and electric distribution technology limited the spread of businesses; these businesses had to remain close to power stations because there was no way to "transport" electricity. However, technology advances in both electric transmission and distribution systems as well as power station design facilitated the growth of industry in cities all across America.

From a geographic perspective, these technology advances shifted populations from northeastern urban centers to all fifty states, which helped many states reap tremendous economic benefits from the influx of new industry. Figures 1-1 and 1-2 illustrate this population shift.

The introduction of electricity to manufacturing processes revolutionized both factory designs and production methods. For

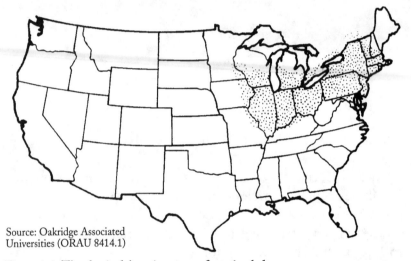

Source: Oakridge Associated Universities (ORAU 8414.1)

Figure 1-1. The classical American manufacturing belt.

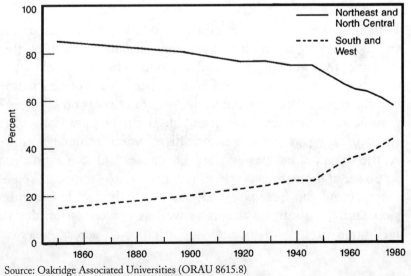

Source: Oakridge Associated Universities (ORAU 8615.8)

Figure 1-2. Employment in manufacturing establishments, by region, 1949 to 1977.

example, by substituting electric motors for complicated belt-driven machine shop systems, a business could physically and technologically reorganize its manufacturing process to achieve greater efficiencies and economic benefits. In the twentieth century, electricity has been used to develop and power some of the most basic technological developments, including the analog computer, the cathode-ray oscillograph, and the impulse voltage generator.

Interestingly, consumers do not always recognize the link between these technological developments, electricity, and the industry that provides this service. They are even less likely to make this association for more advanced electric applications such as robotics and computer-aided production. Researchers create new generations of these products regularly and simultaneously the electroprocesses used to develop them evolve. Yet consumers still fail to equate electricity with this technological development.

As Figure 1-3 illustrates, electricity is recognized as the most significant technical advancement of all time. However, in addition to not recognizing the link between electricity and technology, consumers also fail to see that electricity performs many tasks more efficiently and economically than alternative power sources. The gas industry has mounted a very successful public relations campaign to tout the cleanliness and efficiency of its product. Yet, electric technologies for numerous applications have evolved dramatically in the past decade to meet or exceed these standards. In many of these applications, electric equipment is now on par or beginning to exceed the efficiencies of gas powered equipment. Advances in heat pump technologies, for example, have resulted in efficient and cost-effective electric applications for different needs, such as air, ground source, and water loop heat pumps. Also, high efficiency space cooling equipment now exists for both air conditioners and water chillers. These products produce efficiencies that are up to 50 percent greater than equipment with average efficiencies.

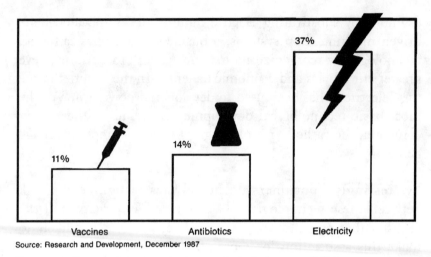

Figure 1-3. Survey of the most signicant technical advances of all time.

Industries are discovering electricity's greater efficiency and added benefits. Electric impulse drying of paper products yields a superior product over gas-fired kiln drying. With the same pulp input, this process increases tensile strength and density and yields a "premium" product for high quality printing applications. Because of this product's enhanced quality, it commands a "value-added" price premium. Electricity also offers purely economic benefits to sugar manufacturers. In this industry, 60 cents worth of electricity performs the same sugar refining task as $3.84 worth of steam produced from gas.

The outside skin of the Defense Department's "stealth bomber" is made from thermoplastic composites. Originally manufactured in gas-fired autoclaves, thermoplastic composites are graphite-based lightweight materials used in the aerospace industry as well as commercial aircraft production. Recent electric technology advances show that by replacing the gas-fired autoclave production method with electric infrared heating, manufacturers can reduce the labor requirements of this process by 60 percent–70 percent, and produce materials that are 9 percent stronger and 5 percent stiffer.

A Historical Perspective: Utility Marketing Efforts

To remind customers of electricity's multifaceted value, the electric utility industry must give this intangible product a form, shape, history, and purpose. Basically, it must readopt one of Thomas Edison's first marketing strategies: make electricity tangible to customers. Electric utility marketing efforts directly affect customers' perceptions of electricity and its value.

Throughout the electric utility industry's history, its marketing strategies and tactics have evolved dramatically; actually, they have come almost full circle since its birth. These efforts have gone through five distinct phases, each characterized by different industry views of its customers.

Phase 1: Introduce the product

In the early days, electric utilities concentrated their marketing efforts on introducing customers to this new product—electricity—and its various uses. Utility representatives went directly to customers to demonstrate what electricity could do for them: light and heat a home, power an iron, and run a sewing machine. Thomas Edison believed that electric utilities had to show customers electricity's tangible benefits, because electricity itself was so intangible to the average person.

Phase 2: The customer becomes a "meter"

As the years passed and electricity became a necessity of life to most Americans, utilities moved away from marketing electricity's uses and concentrated on billing customers. They began selling kilowatt-hours of electricity and defining customers as "meters," or units of sale. Utility sales statistics measured only "numbers of meters set." Since electricity sales climbed yearly, the industry felt no pressure to broaden its sales efforts to include marketing and customer outreach functions.

Phase 3: Boost sales by adding load

Beginning in the 1960s, utilities wanted to increase the number of electricity's end uses to boost electricity sales. During this third phase, utilities' primary goal was to increase sales by encouraging customers to add appliances. Beyond that, they expressed little interest in which customers actually added appliances or how they operated them.

Phase 4: Prices rise and conflicts begin

The Arab Oil Embargo of the early 1970s prompted the shift to the fourth phase. Rapid increases in electricity prices startled customers, who began expressing their displeasure through very vocal, public criticism of utilities. In addition, the mounting costs of building nuclear power plants added to this growing debate, since utilities planned to pass these costs on to ratepaying customers.

Phase 5: Competition and customers

The growing realization in the mid–1980s that electric utilities faced viable competition for their customers pushed the industry into the fifth phase. The concepts of "marketing" were dusted off and reexamined. Utilities slowly began listening to customers again. They learned they had to offer a superior product to keep customers from switching to competing energy options.

Each of these five phases sent customers different messages—about electricity, its uses, its costs, and the utilities themselves. Accordingly, customer perceptions about this product have not necessarily remained consistent over the years. The last three phases, in particular, represent drastic changes in the ways utilities relate to their customers. Unfortunately, the turbulence and antagonism that has characterized many recent customer–utility relationships often negatively impacts customers' perceptions of electricity and its value to them.

The way electric utilities work through this fifth phase will determine the success of the industry. Today's U.S. electric utility industry faces several complex marketplace issues—environmental, economic, political, and competitive—that affect customers' perceptions of electricity's value. The resolution of these issues will shape current and future utility initiatives, regulatory actions, and the ultimate design of customer programs and services.

The complexity of these issues (and the sporadic way they are often publicized) contributes to the promotion of some general misinformation. Views such as "electricity is too expensive," "electricity is bad," "electricity is scarce," and "converting fossil fuels to electricity is inefficient" are often expressed publicly. While there is little or no truth to these maxims, each undermines customers' opinions about electricity's value. For example, electric utilities must demonstrate that while electricity prices have increased significantly in the past fifteen years, electricity still offers superior productivity, efficiency, and power quality when compared with fossil fuels.

The Environment, Politics, and Competition Affect Perceptions of Value

The most comprehensive issue utilities face is "the environment." Customers' awareness about the environmental impacts of generating plants (e.g., nuclear, coal) increases daily and their concerns directly impact their perception of electricity's value. Consequently, their concerns garner an ever-growing amount of media attention, which, in turn, prods politicians and regulators to become involved.

Some environmental issues like global warming, acid rain, and the need to phase out chlorofluorocarbons (CFCs) affect electric utilities nationwide. Worries about the impacts of coal-fired plants on global warming and acid rain have forced utilities to quickly adopt load reduction strategies, because most regulators discourage or virtually prohibit programs that increase electricity sales. As a result, to continue supplying their customers' electricity demands, utilities must manage their current supply-side resources more effectively by improving their overall efficiency levels.

Manufacturers use products such as CFCs and hydrofluorocarbons (HCFCs) to improve energy efficiency levels in many of today's electric refrigeration systems. Unfortunately, these products also pollute the environment; a recent review of the Montreal Protocol in London, England and amendments to the Clean Air Act will phase out CFCs by the year 2000 and HCFCs by the year 2030. To counter these changes, the utility industry must become involved in efforts to develop cost-effective alternatives which protect electric market share and continue to offer customers equivalent, energy efficient comfort levels.

Some environmental issues, while they may have indirect global effects, are more geographically and utility specific. Air quality problems are no longer confined to major metropolitan centers such as Los Angeles and New York City. In fact, in cities throughout the country, public concerns about air quality have prompted an increasing amount of legislation. The electric utility industry is working to offer several solutions to this very real problem. Chapter 10 addresses just one; it discusses the industry's efforts to develop and deploy electric vehicles.

Another issue, public perceptions about the effects of static electric and magnetic fields found around appliances and other devices, also creates substantial perception problems for the industry. If customers continue to feel uncomfortable about the possible negative effects of using electric equipment, utilities will face enormous customer service costs (in responding to complaints) as well as lose customers to fossil-fuel-based energy services. Utilities must assess the actual electric and magnetic field impacts of utilization devices and take the lead in developing technology to counter any adverse impacts.

Increasing competition magnifies the "perception" problem—

customers who have choices often let their perceptions enter their decision making processes. Electric utilities often face competition from other utilities in their areas (e.g., gas or another electric supplier), nonutility generation, and foreign competitors. Even when U.S. manufacturers move offshore, they too are switching to "the competition." In addition, increasing transmission access and the rapid growth of nonutility generation threatens even secure markets.

Recently, utilities and the Electric Power Research Institute (EPRI) have uncovered a wealth of information about residential and commercial customers, including how they make energy use decisions, which energy attributes are important to them, and what they want from energy service. Using this information, utilities can develop energy programs and services targeted at specific customer groups. The key, however, is to convince customers of the *value* of these energy services:

"What the customer thinks he is buying, what he considers value, is decisive—it determines what a business is, what it produces and whether it will prosper. And what the customer buys and considers value is never a product. It is always utility, what a product or service does for him."[1-1]

What Can Utilities Do To Enhance Electricity's Value To Customers?

Strategically, the electric utility industry must take control of its own destiny and mold customers' perceptions: it must successfully educate customers about electricity's value and the role it plays in their daily lives. Electricity's value comes from its utility—the basic needs it fills (comfort from a warm room on a cold day), the tasks it performs (cooking food), the services it renders (money from an automatic teller machine).

As illustrated in Table 1-1, utilities can enhance electricity's value in three ways:

- Increase customers' perceptions of electricity's value
- Focus on electricity's own technical and economic attributes
- Incorporate the concept of "enhanced value" into all planning processes

Table 1-1
Three Ways to Enhance Value

Value Enhancement	Outcome
Increase customer perceptions of value	Increasing perception is everything: real value may not be meaningful
Enhance electricity's technical and economic attributes	Deliver more real value
Make value an integral part of meeting customer needs	Both increase perception and deliver value

First, utilities can enhance electricity's value simply by increasing customers' *perceptions* of its value. They can discover what customers want and need and then show them how electricity meets these requirements. Utilities must recognize that customers do not perceive a need for electricity specifically—they *do* think they need heat, light, stereo systems, and "quick cash" machines, however. (Similarly, customers failed to perceive that they needed, or would use, personal computers until they became readily available and accessible.) If they increase customer satisfaction, utilities can also enhance their own public images, which, in turn, impacts perceptions about electricity.

Second, utilities can enhance the *technical capability* of electricity to do work. As end-use products (appliances/devices) become more functional and meet customer needs more effectively, they become indispensable to customers' everyday lives. Each successive generation of electric powered products should be able to "do it quicker, sooner, and more effectively." Utilities must link these improvements with electricity's role in facilitating them.

In addition, customers will not recognize the full potential of electricity's value until utilities deploy the most cost-effective programs/services. Electricity's *economic value* will improve as utilities design and promote more effective demand-side management (DSM) programs that meet actual customer needs. Utilities that match customer needs to the right energy service programs will design and implement the most economic, cost-effective programs.

Finally, utilities that want to enhance electricity's value must consciously do so—they must incorporate this goal into all of their planning processes. Traditional utility planning focused solely on supply-side issues. This traditional process assumed stable and predictable load growth; "the customer" received very little consideration in this process. As the utility marketplace changed, utility planning processes had to evolve. Accordingly, utilities began adopting processes that focused more on demand-side issues, such as least-cost planning. However, traditional least-cost planning procedures concentrated on economic motivations and neglected other factors that may prompt customers to purchase utility services—such as perceptions of value. An integrated planning approach—both in resource and program planning—incorporates both economic and noneconomic factors. Chapter 13 discusses these planning processes in greater detail.

Some Current Examples of Value-Based Programs/Services

Many electric utilities are turning to a value-based philosophy when designing, implementing, and promoting programs and services. While many utilities may not have labeled their actions as such, the evidence mounts, including the manner in which they develop and market DSM programs and services. These options include:

- *Contract bidding.* Utilities request proposals from small power producers.
- *Backup power.* Utilities build and operate backup power systems for customers.
- *Investments in alternative technologies.* Utilities (solely or in partnership) invest in alternative supplies such as geo-thermal plants or wind farms.
- *Architectural design assistance.* The utility arranges appointments with architects for customers who are planning new construction projects. These architects help the customer design energy efficient buildings.
- *Economic development rates.* Currently offered by utilities serving depressed or new development areas, these low rates encourage production or investment in the area. A similar rate is sometimes offered to plants in industries facing potential shutdown due to foreign competition.
- *Prepaid electric service.* Customers can purchase magnetically encoded cards that remotely operate meters installed in homes. The home meters are equipped with a digital display of information, including current rate of energy use, cost of cumulative service, and the amount of energy used on previous days. Warnings notify customers at preset fill levels. Utilities that have implemented this program found that the use of this device has cut meter reading expenses by 50 percent. In addition, credit risks are lowered and the customer does not have to worry about unexpectedly high bills or a loss of service due to nonpayment of bills.
- *Spot gas credits.* Credits are offered to customers who buy spot gas and allow the utility to burn it in its gas-fired generators.
- *Five year rate contracts.* Each year's rate increase is specified in advance. This service was offered to one utility's major industrial customers because these customers feared future rate increases due to the construction of a nuclear power plant.
- *Regional rate option.* Utilities offer industrial customers the option of paying an average of twenty-four regional utili-

ties' rates versus the serving utility's rate.
- *Energy service company incentives.* The utility provides incentives to energy service companies that enter into performance contracts with industrial or commercial customers having a demand of at least 500 kW.
- *Real-time pricing.* Utilities announce tariffs weeks, days, or even hours before use; these tariffs apply for a short time only (in some cases separate rates for each hour of the day are offered).

Finally, a survey of current utility activities shows that many utilities now promote new technologies in a variety of innovative ways. Many utilities now sell, lease, install, and/or maintain equipment for a variety of end uses. (This is a controversial practice in many states, since contractors and others believe utilities are usurping their roles.) For example, Red River Valley Electric sells and installs heat pump water heaters, Taunton Municipal Lighting leases fluorescent light bulbs, Alabama Power leases and sells power conditioning equipment, Mississippi Power loans customers electric cooking equipment, and Wisconsin Electric Power offers a range of consulting services to its customers. The common denominator for all of these utilities: devise unique ways to promote electricity's use and meet customers' needs in the process.

A Few Overall Suggestions for Improving Value-Based Marketing Efforts

Electric utilities have reintroduced themselves to their customers during the past few years. They are learning how their customers have changed, what customers want and need from them, and what they, as utilities, can do to serve customers efficiently and cost-effectively. Utilities that learn these lessons will succeed in today's competitive energy marketplace; those that continue to view customers as mere inconveniences will face losing their businesses to more customer-oriented energy suppliers.

Slogans of the 1930s and 1940s ("the Customer is King" and "the Customer is **Always** Right") have been resurrected by a variety of companies. These organizations, including Federal Express, United Parcel Service, AT&T, Nordstrom department stores, and Apple Computer, all recognize the importance of convincing customers that they are valued, appreciated, and well served. Tom Peters, international management expert, puts it another way: "...look out elephants, the gazelles are taking over!" He reminds all businesses that companies who anticipate and respond quickly to customer needs will win in a competitive business marketplace. *Electric utilities already meet customer needs quickly* (to the nanosecond, actually). They just need to remind customers of this fact. After all, Federal Express and United Parcel Service deliver, but not at light speed!

Utilities must do everything they can to help their customers succeed: know them, help them, work to deserve their trust. "Know them" means relate to them as people—not as meters or units of revenue. Electric utility customers are renters, home owners, small business owners, farmers, and manufacturers. They have problems and concerns that utilities must understand; in addition, utilities must discover which factors affect energy use decisions. Utilities must find out *what* these customers want, *how* they want it, *when* they want it, *why* they want it, and *how much* they are willing to pay—or not pay. These findings (and what utilities do with them) impact perceptions of value.

"Help them" refers not just to providing electricity at competitive prices. It means responding quickly to their needs in a skillful manner. In many cases, utilities should anticipate customers' needs and provide a service *before* the customer knows he/she needs it. Too often, utilities are caught responding to problems—at this point, they face losing customers.

"Earn their trust" means be credible—do not hustle the customer or play "government baseball." If, after careful analysis, an alternative to electricity turns out to offer a customer more benefits, admit it, let the customer choose his/her preference, and help make the conversion, if necessary. Utilities that maintain their integrity even in a difficult business situation will earn their customers' respect. (A utility plays "government baseball" when the rules get changed in the middle of a game. Utilities that encourage customers to install special equipment in exchange for a beneficial rate—and then rescind that rate a year later—play government baseball. Once played, this game only leads to dissatisfied customers who will mistrust every utility attempt to provide future assistance.)

Remember, customers value many different attributes of a product, some of which have nothing to do with the product itself. Electric utilities that understand the importance of increasing electricity's value to customers will:

- Meet their customers' needs as well as their own.
- Market their programs and services to customers more effectively.
- Maximize profitability and competitiveness.

The remaining chapter in this section addresses the different attributes of electricity. Section II focuses on electricity's impacts on the environment, national security, family life (including women's roles in U.S. society), and economic. Section III discusses electrification's impacts on industry; practical case studies give concrete examples of electricity's growing value to several industries. Finally, Section IV offers strategies for the future. Its chapters include practical "how to" advice to utilities. They explain how value-oriented utilities can adjust their planning processes, meet their customers' needs, and create effective marketing programs.

Customers are complex human beings, motivated to make purchases by a wide variety of tangible and intangible factors. If they do not understand or cannot perceive the value of a product—no matter how wonderful or useful that product is—they will purchase a competitor's product. This book as well as many other tools can help electric utilities understand why the concept of value is so important as well as how to communicate this concept to customers.

References:

1-1 *Management*, Peter Drucker, p. 84-86, Harper & Row: NY, 1973.

CHAPTER 2
Electricity's Attributes

Introduction

Electricity is a uniquely valuable form of energy, offering unmatched precision and control in application as well as efficiency. It offers unrivaled environmental benefits when compared with other energy options. And finally, electricity provides individuals with a clean comfortable supply of energy. Because of these unique attributes, new electric appliances and devices require less total resources than comparable natural gas or oil-fired systems.

Electricity's utility is so diverse. Certain energy forms can meet one need more efficiently than electricity, but these forms are extremely limited in their range of application. Only one energy form—electricity—can meet all of a customer's energy needs (comfort, convenience, appearance) as well as facilitate the achievement of other needs (medical diagnostics, money from automatic teller machines, personal computers). Electricity is extraordinarily unique in its ability to deliver packages of concentrated, precisely controlled energy and information efficiently to any point.

In addition, electricity can help alleviate many of the concerns facing this country and the world (e.g., environmental problems, limited resources and the spiraling costs for obtaining them). In fact, it is uniquely suited for this critical task:

- It is available from various sources at a reasonable cost.
- Its versatility allows it to be readily converted into easily and efficiently usable forms.
- Its efficiency at the end use comes from its versatility.

Electricity's efficiency at the end use is critical: 1) for future energy, resource, and environmental "conservation"; and 2) to rebut the incorrect perception held by some that electricity is an "inappropriate" power source due to its "low efficiency."

This chapter discusses several attributes of electricity in greater detail and then presents a wide variety of electrotechnology applications in homes, businesses, and industries.

Attributes of Electricity (2-1)

Electricity offers society more than just improved energy efficiency. It also has greater "form value" than any other energy source: form value affords technical innovation with enormous potential for economic efficiency. Form value encompasses three dimensions: technical, economic, and resource uses.

Technical

Electric powered equipment offers significant advantages when compared with conventional (e.g. thermal) counterparts, including reduced energy use, increased productivity, and improved product quality, compactness, and environmental cleanliness. Electricity's technical attributes include: electrical phenomena, input energy density, volumetric energy deposition, controllability, and synergistic combinations.

Electrical phenomena

Residential appliances, building energy systems, and industrial processes frequently involve the interaction of energy and matter to modify materials, pump refrigerants or fluids, or to transform them from one form to another. Three types of electrical phenomena can be involved in these transformations: electromotive, electrothermal, and electrolytic. These phenomena are all unique to electricity as an energy form and contribute to its form value.

- Electromotive phenomena occur when mechanical motion is produced using electricity. The electric motor represents the most prominent example of electromotive phenomena, accounting for nearly 80 percent of the electricity consumed in process industries. The electric motor is by far the most efficient and effective source of motive power: users can obtain efficiencies of over 90 percent with this device. The electric motor is used in a wide variety of applications, including as "prime movers" (pumps, fans, and compressors), materials processors, and handlers. In materials processing, for example, electricity can exert force without physical contact, permitting precise manufacturing of metal parts by rapidly accelerating them against a form.
- Electrothermal phenomena employ electricity to produce heat, which, in turn, facilitates a physical or chemical change. Bulk processing industries (e.g. chemical, primary metal, stone, clay, glass, paper, and petroleum) transform material from one physical or chemical state to another. For example, three techniques exist. In the simplest method, direct ohmic dissipation, an electric current is passed through the material by physically attaching electrodes. The second method, electromagnetic induction, heats conducting materials without direct physical contact. The material to be heated is placed in proximity to a coil carrying an alternative current; the fluctuating magnetic field produced by the coil induces eddy currents in the material, which are dissipated to produce heat. Finally,

certain nonconducting materials can be heated dielectrically. This process occurs when a material containing polar molecules (e.g., water) is placed in a rapidly alternating electrical field. Dielectric losses produce heat, as, for example, in a microwave oven.
- Electrolytic processes bring about chemical change through the direct use of electricity. Electrolytic phenomena occur at the molecular and atomic levels. The earliest (and still most common) industrial applications of electrolytic processes occur in the chemical industry. Electrolysis is best known for its use in the production of such basic materials as aluminum and chlorine.

Input energy density

In combustion processes using chemical fuels (e.g., oil and gas), the maximum achievable temperature is thermodynamically limited to the "adiabatic flame temperature," a practical limit of about 3000°F for fossil fuels burned in air. When heating material electrically, there is no inherent thermodynamic limit on the temperature. Typical temperatures of 10,000°F and higher are achieved routinely in arc-produced plasmas and much higher temperatures are technically feasible.

Volumetric energy deposition

Electrothermal phenomena are volumetric (i.e., generating heat within the material itself). When using fossil fuels to heat material, heat is usually imposed at the surface by radiation and convection. This method is inherently slow and inefficient. With induction heating, electrical energy is deposited directly within the material; thus, a process can be reduced to several minutes or less.

Volumetric heating can also be used to dry moist materials with microwave or radio frequency radiation. In conventional drying, heat must diffuse into the material from the surface, while moisture diffuses out, a slow process, since most materials of interest are poor thermal conductors. If drying is accelerated too much by intensifying the rate of heating, overdrying of the surface can occur, causing cracking and degradation of the product. Dielectric heating largely eliminates this problem by greatly increasing drying rates. Thus, this process improves overall productivity, product yield, and product quality.

Controllability

Electricity is often referred to as an "orderly" form of energy, in contrast to thermal energy, which is random in nature. This reference means that electrical processes can be controlled much more precisely than thermal processes. Since electricity has no inertia, an industry can instantly vary energy input in response to process conditions, such as material temperature, moisture content, or chemical composition as well as accurately maintain a desired state. Lasers and electron beams can be focused on a work surface to produce energy densities a million times more intense than an oxyacetylene torch. The focal points of these high intensity energy sources can be rapidly scanned with computer-controlled mirrors or magnetic fields to deposit energy exactly where it is needed. This focusing capability offers a tremendous advantage, for example, in heat-treating a part precisely at points of maximum wear, thereby eliminating the need to heat and cool the entire object. Electrolytic processes impart energy directly to ionic species to produce molecular separation or selectively induce chemical reactions.

Synergistic combinations

In some processes, electrolytic, electrothermal, and electromotive effects combine in an advantageous way. In the Hall-Heroult process for reducing alumina to aluminum, ohmic heating helps keep the cryolite bath in the molten state, while electrolysis causes pure aluminum to separate out and collect at the cathode. In a coreless induction melting furnace, electromagnetic induction heats and melts the charge while at the same time inducing a strong electromotive stirring action, which enhances heat transfer to the solid material and greatly improves homogeneity of the melt. The latter effect is especially important in the production of high-alloy castings and can be a major determining factor in the choice of induction melting over alternative methods.

Economic

Economic attributes include: fixed cost, flexibility of raw material base, and product quality and yield.

Fixed cost

High energy density and precise control typically result in increased production rates, which, in most cases, reduce fixed costs per unit of product. With, for example, faster throughput, components such as labor, overhead, and interest on capital are spread over a larger production volume. Thus, even when the cost premium of electricity increases the energy cost per unit, total production cost per unit may, in fact, remain lower.

Electrical process equipment is typically economical in smaller unit sizes than combustion equipment, since it does not require fuel handling and environmental control. Electrical production technology can give rise to an "economy of reduced scale," in that smaller equipment requires less space, and industries can decentralize facilities and site them in proximity to diverse raw material sources and product markets. This flexibility, in turn, carries with it great economic benefit.

Flexibility of raw material base

The high energy intensity and precise control offered by electrotechnologies permit a greater degree of flexibility with regard to raw material resources than do most conventional processes. Arc furnaces, for example, can utilize either scrap or direct-reduced iron as a basic resource with little or no process modification.

Product quality and yield

In overall production economics, product quality and yield are critically important. Electrical processes typically provide improvements in both areas.

Resource Use

Resource use attributes include: flexibility of fuel supply, domestic resource balance of payments and national security, environmental, and energy consumption.

Flexibility of fuel supply

Combustion-based processes are highly dependent on the availability of specific fuel sources, since combustion equipment in general is not very adaptable to changes in fuel type. The shift to electrically-based processes assigns responsibility for fuel choices to the electric utility, which can optimize fuel diversity. In addition, efficient energy conversion is a utility's primary concern. Thus, over the long term, basic manufacturing processes or business activities can remain essentially the same while the utilities respond to the changing situation in primary fuel markets.

Domestic resource balance of payments and national security

Total resource requirements, and therefore the need for imported fossil fuels, declines with the increased use of electricity due to the overall efficiency of electrically-based systems. As these imports decline, the impact on our balance of payments declines as do the risks in protecting our national security.

Environmental

Electric processes and systems are undisputedly the most environmentally benign at the point of end use. Recently it has also become evident that the application of electrotechnologies can mitigate what would otherwise be adverse environmental impacts in transportation, industrial, municipal, and medical waste applications. Even when converting raw resources to produce it, electricity becomes the environmentally preferred energy form, due to its efficiency at the point of end use.

Energy consumption

Initially, one might think that primary energy consumption should always be higher for electrical processes than with conventionally fueled systems. This perception is strengthened especially in light of the roughly three-to-one conversion ratio of fuel to electricity at the power plant. In most cases, however, the opposite is true as a result of exploiting the form value advantages of electricity. Historically, energy intensity has declined as electricity has increased.

Total Resource Efficiency

To compare and analyze the total resource efficiencies of electrically-based with fossil-fueled end uses, analysts should consider the total energy "system" from the raw resource through conversion, delivery, and utilization. In comparing fossil fuel with electricity use, analysts must consider all losses of conversion and delivery as well as the gains from leveraging alternative energy sources (e.g., a heat pump's use of solar heat or advanced technologies such as information technologies in homes).

The system illustrated in Figure 2-1 depicts one unit of energy being derived from an oil or gas well. In an electrical system, it is converted to electricity at an assumed heat rate of 11,000 Btu/kWh. This conversion results in 0.31 units of electricity from the original 1.0 unit of fossil fuel. Applying a steady-state loss of 8 percent in transmission results in a delivery of 0.29 units of energy to the electric end use.

Typically, gas transmission and distribution systems lose 10 percent of their energy in pumping and leakage. This results in

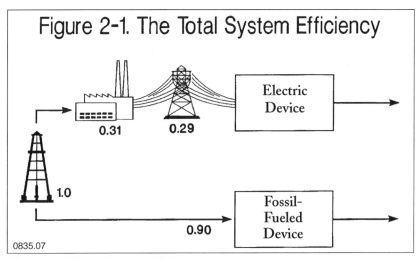

Figure 2-1. The total system efficiency.

0.90 units of energy being delivered to the gas end use from the original 1.0 available at the well head. This case includes the losses in refining, transportation, and dispensing.

Practical Examples

Several examples summarized below describe the benefits of examining total energy system efficiency; they reveal the breadth of the benefits of the wider use of electricity.

Freeze concentration

Process industry opportunities to enhance electricity's value by increasing energy efficiency are truly promising: recent advances in freeze concentration provide one exciting example. Freeze concentration technologies separate substances in crystalline form at substantial savings in cost and energy. These technologies have a wide range of potential applications, from the preparation of food and chemicals to the treatment of wastewater. For example, the dairy industry is the largest user of energy for freeze concentration of any of the food industries. The equipment now used is generally antiquated—most typically employing evaporators. This technology is not nearly as efficient as freeze concentration and uses large supplies of fossil fuel. The dairy industry has exhibited interest in freeze concentration to replace thermal evaporation for several years. (2-2)

As seen in Figure 2-2, freeze concentration uses an electrically-based vapor compression system to freeze out the water in a product. Freezing requires a total of 144 Btu/lb of water, or 114 Btu/lb with heat recovery. Evaporating or boiling off water requires considerably more energy—in this case, a net of 700 Btu/lb with heat recovery. When the total systems are compared, the electric-based process has at least twice the overall efficiency as the natural-gas-based process.

In addition, applying freeze concentration in the dairy industry not only yields superior quality products, but based on only 10 percent market penetration of this technology, the utilization of freeze concentration would save 3.4×10^{12} Btu/year of fossil fuel. This reduction in the use of fossil fuel would significantly reduce pollution and foreign oil dependence. In addition, freeze concentration technology operates at lower temperatures which reduces microbiological and enzymatic activity, thereby allowing better food quality, equipment utilization, and lower cleaning costs.

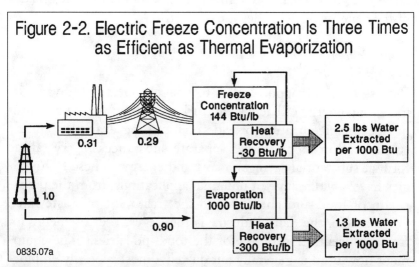

Figure 2-2. Electric freeze concentration is three times as efficient as thermal evaporization.

Electric heat pump

One of the single greatest electric powered technological advancements (outside of computers and medical electronics) is the electric heat pump. Chapter 3 discusses this electrotechnology in greater detail. The most significant benefits of the heat pump are disguised by its very name "heat pump." It is fundamentally the most efficient solar machine available. The heat pump uses the heat in outside air or in the ground to transfer the sun's energy to buildings. This process yields tremendous system effectiveness. Electric heat pumps are available today with a steady-state efficiency almost four times that of an advanced gas-fired condensing furnace. One such heat pump now on the market is Carrier's "HydroTech 2000," the result of a six-year multi-million dollar research program sponsored jointly by EPRI and Carrier. Figure 2-3 depicts the Carrier unit in comparison with a pulse combustion furnace. Because of its use of electronic variable speed compressors and blower drives, a patented refrigerant circuit for year-round integrated water heating and a novel coil defrosting method, it is the most efficient space conditioning and water heating system available in the world today.

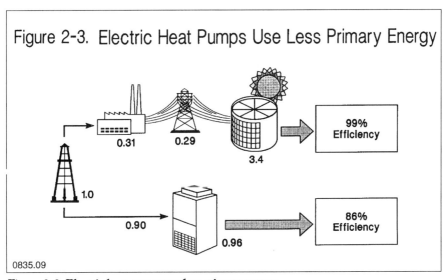

Figure 2-3. Electric heat pumps use less primary energy.

Using Philadelphia as a test site, EPRI recently conducted a more detailed study of the resource energy requirements of this electric heat pump compared with those of a gas pulse combustion condensing furnace. Consistent with the conclusions in Figure 2-3, the study results show that gas furnace/electric air conditioner combination consumes 19 percent more resource energy (gas) than the electric heat pump.

Electric transportation

Electric transportation presents an excellent example of an electrotechnology application that harbors tremendous environmental benefits. The inadequacies of the country's transportation systems are all too apparent. Growing traffic congestion, insufficient public transit, and the noise, pollution, and the frustrations they create are frequent sources of feature stories and commentaries and most recently of regulatory and congressional hearings and regulations. Virtually no person living or working in a major metropolitan area remains untouched by this situation, which is steadily growing worse.

For many years, the electric power industry has worked with major automotive manufacturers to develop reliable, practical electric vehicles (EVs) that meet industry and consumer standards. This effort has produced two electric vans, the GM Electric G-Van and the Chrysler TEVan minivan; both are slated for use in commercial service fleets within the next few years. Figure 2-4 compares the energy use of one of these near-term electric vans with their gasoline-powered counterparts. The figure traces the conversion of one barrel of oil through the electrical system into an electric vehicle and, in contrast, through refinery and fuel delivery systems in a gasoline powered van. Due to the inherent efficiencies of the electric motor and refinery/distribution losses, even this fledgling technology is 60 percent more efficient than its gasoline counterpart. [2-3]

For gasoline vehicles, sources of emissions include both tailpipe emissions and the emissions associated with refinery operations

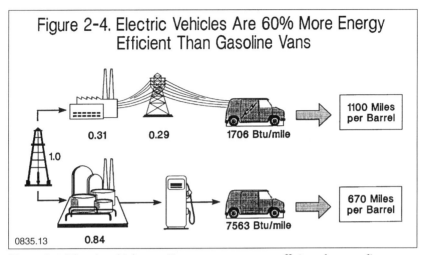

Figure 2-4. Electric vehicles are 60 percent more energy efficient than gasoline vans.

and fuel distribution to deliver gasoline to the vehicle. Electric vehicles do not emit any pollutants during operation. All emissions associated with EVs come from the electric power plants that provide electricity to the vehicles and are totally dependent on the generation mix used. Different types of power plants display very different emission characteristics. Future fossil-fueled power plants will be considerably cleaner than the fossil-fueled power plants that are operating today.

Chapter 10 discusses electric transportation in greater detail.

Information technologies

Increasingly, America's economy is becoming service-based. This shift has dramatically affected the use of information technologies in today's workplace. The growing need to manage and disseminate information has led to innumerable advances in electric powered technologies. Concurrently, major changes in workers' job responsibilities have led to new stresses.

The need for psychological privacy to allow the eye and mind to identify a sense of private place may inhibit further dramatic increases in worker density in existing office environments. Some futurists have predicted that by the year 2000 we will see a move away from office-based employee teams toward the "electronic cottage." They envision the electronic cottage as a cozy room where all electronic gadgets would put the at-home office worker in kind of a command console—an intriguing idea that many workers are testing today. Analysts expect the "cottage" will probably not see significant use in this century, but may change energy use for these technologies by 2010.

Others have advanced the theory that the office of the future would contain fewer objects, in essence, be friendlier than a collection of all currently known office technologies. This theory could only be possible by integrating some technologies, perhaps combining the FAX machine, telephone, photocopier, and personal computer into one machine. Today's cutting edge appliances are beginning to reveal this trend (for example, laser printers that now have FAX and copy machine capabilities).

The movement of information presents the key to energy service needs in the commercial sector. Electrons will carry the information. These electrons now normally supplied by electric utilities to just provide artificial illumination, motive power, cooling systems, and efficient heat pump systems, will increasingly energize information technologies. (2-4)

"Telecommuting," or the use of electronic systems instead of commuting, is increasing in popularity due to the growth in productivity which results. The FAX machine represents one interesting example of the expanding, efficient use of electricity-based technologies to move information. Figure 2-5 depicts the use of a FAX machine in relation to an express courier service. In this conservative example, FAX can send and deliver 177,000 pages per barrel of oil as compared to 25,200 pages per barrel by a courier service.

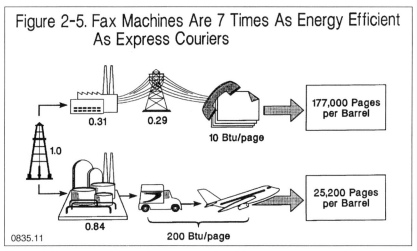

Figure 2-5. Fax machines are 7 times as energy efficient as express couriers.

New Opportunities for Electrotechnologies

The examples in the preceding section compare the total resource efficiencies of several electric powered and fossil-fueled technologies. As Figures 2-1 through 2-5 so clearly illustrate, electric technologies offer superior efficiencies. In fact, when these existing applications are combined with new opportunities for electrotechnologies, the net effects in electricity consumption are substantial.

Table 2-1 depicts the likely distribution of consumption increases offered by the existing and new opportunities for electricity use, by sector. Somewhat surprising is the large number of new opportunities—often referred to as beneficial electrification—in the residential and commercial sectors. These sectors' energy uses are driven by space conditioning and information systems.

Table 2-1
Consumption Increases Expected
From Beneficial Electrification, by Sector
(Billion kWh Increase Over 1990)

Sector	2000	2010
Industrial	66	128
Municipal	11	18
Commercial	167	267
Residential	145	256
Transportation	6	34
Total	395	703

The net effect of the growth in electricity use produced by beneficial electrification could total as much as 703 billion kWh in the year 2010. Due to electricity's excellent "form value," EPRI researchers estimate this increase in electricity use will lead to a net reduction in total energy use of seven quads and a reduction in carbon dioxide emissions of over 440 million tons. Wiser electricity use (utility DSM programs coupled with regulatory incentives) could reduce consumption by 460 billion kWh in the year 2010 with reductions in energy use of 4.6 quads and CO_2 of 330 million tons.

If we compare the impacts of beneficial electrification achieved through wider use of electricity with those achieved through wiser use, then we find that both measures will reduce CO_2 emissions, although wider use obviously increases electricity use and wiser use reduces it. Analysts estimate combination of the wiser and wider use will reduce total energy use by over eleven quads and CO_2 emissions by 770 million tons in the year 2010.

This estimate projects a total electricity demand of nearly 4246 billion kWh in the year 2010. This number represents a net increase of 201 billion kWh due to the consideration of opportunities for either efficiency or beneficial electrification. That is, the expanded use of electricity due to its wider use will likely outweigh our best efforts to use electricity even more wisely. And the fraction of total energy served by electricity will likely continue to increase as electricity continues to be substituted for less efficient and productive energy forms.

Some who argue for energy conservation discount the idea of "wider" use. However, those who speak about saving energy must consider this approach because of the environmental and efficiency improvements it offers. Only through the expanded use of electricity (wider use) coupled with aggressive electricity end-use efficiency programs (wiser use) can we truly save energy.

The remainder of this chapter offers practical examples of applications for electrotechnologies—these examples illustrate both wider and wiser uses for electricity.

Iron and Steel Industry[2-5]

In 1973, the iron and steel industry used 17 percent of manufacturing's total energy demand. Between 1970 and 1982, this industry's total energy use per ton fell by 25 percent, but its electricity consumption, per ton, increased 20 percent. Two major developments in the area of electric technology will cause these growth trends to continue: 1) the increasing use of electric arc furnaces to remelt scrap, and 2) the advent of minimills.

Events in the 1970s reduced electricity's cost, compared with other fuels, and motivated industry members to use electrically-powered processes to melt steel. As a result, the share of raw steel melted electrically grew from 10 percent in 1965 to 20 percent in 1975; it now stands at 35 to 40 percent. The economic reasons for this shift are compelling: first, electric scrap melting requires less than half of the primary energy of steel production from ore. Second, the cost of scrap is less than half of molten pig iron. Third, electric steel production efficiency is 10 percent higher than the more traditional open hearth and basic oxygen processes.

New steel now contains approximately 30 percent scrap, compared with 15 percent in 1970 (Figure 2-6). It is uncertain how far U.S. industry can move in the direction of recycling obsolete steel scrap, but the amount of recoverable scrap in the U.S. is very large (700 million tons) and growing. Even if electric arc steel production using obsolete scrap does not continue to penetrate the market, we can expect use of electricity per kilogram of steel to increase. Processing low grade U.S. ore requires about 250 kWh per ton of raw steel just for ore concentration. In the future, plasma arc technology will likely be used to economically reduce iron ore, and some scrap will be processed cryogenically.

Although the industry still uses electric induction furnaces, particularly to produce specialty steels, a growing amount of today's steel is produced in electric arc furnaces.(2-6) (An induction furnace produces heat by using electrically created magnetic fields, while the arc furnace produces heat by passing an electric current directly through the raw material.) In the electric arc process, workers load steel scrap into a brick-lined, water-cooled steel vessel. They then place a cover containing three large graphite carbon electrodes, typically two feet in diameter, and strike an arc between each pair of electrodes. When the charge reaches a molten state, oxygen is injected beneath the surface of the bath. When the melt is ready, workers pour the molten steel into a

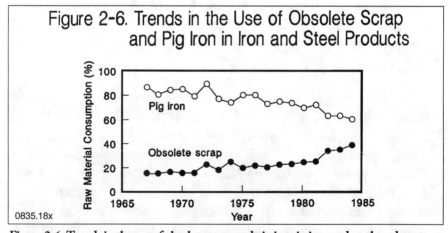

Figure 2-6. Trends in the use of obsolete scrap and pig iron in iron and steel products.

ladle for transfer to casting operations.

The minimill

The switchover from open hearth to basic oxygen processes for steel production created an opportunity for new steel producers to enter the competition. Producers can build small minimills with a relatively modest investment. These mills take advantage of a greater availability of local scrap and, because of their relatively small size, can be located virtually anywhere, thus avoiding the cost problems associated with escalating freight rates.

Steel finishing

Conventional practice for steel casting and finishing operations in integrated steel mills (i.e., mills with facilities for coking coal and smelting iron ore, in contrast to those engaged solely in the remelting of scrap) consists of pouring the steel into large ingots. These ingots are subsequently reheated and, following a series of forming and reheating steps, are made into basic steel products for shipment.

These steel finishing operations represent a substantial part of the overall cost of steelmaking. The costs occur because this process requires several separate operations, and each carries a cost for labor, energy, and capital equipment and contributes to product waste. For example, these operations can produce a ton of raw steel from ore using about eighteen MBtu of primary energy. But, about forty MBtu of energy is expended for each ton of steel product shipped. In addition, only about 70 percent of raw steel production typically ends up as shipped products—producers must recycle the rest. Many modern mills improve these percentages because they make extensive use of continuous casting techniques.

Continuous casting

In this process, a funnel feeds molten steel into a water-cooled oscillating copper mold where the steel solidifies and is continuously withdrawn from the bottom of the mold at a rate of about ten feet per minute. This ribbon of steel is then cut into appropriate lengths (billets) which, in the newest plants, are moved immediately through the rolling mill and emerge shortly thereafter as finished shapes ready for shipment. In this way, 90 percent or more of the molten steel ends up as finished product. This procedure represents an enormous productivity gain compared with traditional practice.

In principle, producers could deliver all steel to finishing operations via continuous casting regardless of the steel production process. In practice, however, continuous casting operations are most easily and economically introduced in conjunction with new steelmaking facilities. These new facilities can match the capacity and design of the steelmaking equipment with the capacity and layout of the finishing section and with the product line planned for the operation. A small, regional minimill fits this specification: it produces a small variety of high-volume simple shapes. In Japan, where more than 70 percent of the steelmaking capacity came on-line after 1963, more than 80 percent of steel production was continuously cast by 1982. In that year the U.S. continuously cast about 25 percent of its steel production.

Energy requirements

Table 2-2 illustrates the energy requirements for the various steps in raw steel production. Table 2-3 lists the energy required, per ton of raw steel produced, for three competing steel processes. (Note that the value given for steel processing using electricity includes the energy used to generate the electricity.) Part of the reason for the large difference in energy requirements is that electrically processed steel comes from scrap, and therefore little energy is required to reduce iron oxide to elemental iron. In this sense, U.S. stockpiles of iron and steel scrap represent a significant source of stored energy.

Table 2-2
Average Energy Requirements for Process Steps
in the Production of Raw Steel (MBtu/ton)

Operation	Energy Requirement*
Ore beneficiation	1.7
Ore transport	0.5
Blast furnace	14.6
Steel production	
Open hearth	4.1
Basic oxygen	1.3
Electric	5.3
Scrap processing and transport	0.6

*Use of electricity counted at 10,600 Btu/kWh.

Table 2-3
Energy Required per Ton of Raw Steel Produced

Steel Process	Total MBtu per ton Raw Steel
Open hearth	14.9
Basic oxygen	15.1
Electric	6.3

Note: By considering the actual amounts of pig iron and raw steel used in each process and the energy associated with those inputs, researchers can compute the total energy required to produce steel by the various methods.

The contrast in energy requirements presents a much more dramatic picture for the minimill with high production efficiency for steel finishing operations. For example, a minimill with continuous casting and electric conductive billet heating might use two to three MBtu per ton in steel finishing and ship in excess of 90 percent of its raw steel production. The industry as a whole, on the other hand, might use an additional twelve MBtu per ton of raw steel in finishing operations, and ship only 70 percent of the raw steel. Thus, the overall production energy is about forty

MBtu per ton of steel shipped from integrated mills with conventional steel furnaces and only about ten MBtu per ton of shipped steel for electric mills with continuous casting.

This substantial reduction in minimills' primary energy use offers distinct benefits for both steel producers and U.S. society. Steel producers that use electric melters pay lower energy bills because much of the fuel displaced by electricity is coking coal and coke, which are as costly as gas or oil. (Coking coal and coke comprise about 40 percent of the energy used to make steel products via the open hearth and basic oxygen processes.) U.S. society benefits from the energy conservation and environmental quality gains.

Figure 2-7 illustrates the impact of the electric mills on overall steel industry energy consumption. As shown, the energy use

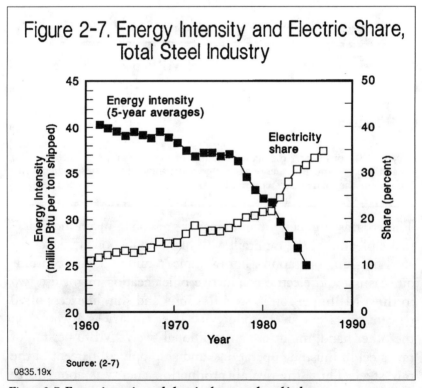

Figure 2-7. Energy intensity and electric share, total steel industry.

intensity for steelmaking held relatively constant at about forty MBtu per ton of steel shipped during the 1960s while the basic oxygen process largely displaced open hearth steelmaking. Beginning in the late 1960s, electric steelmaking began to grow and progressively reduce the average primary energy requirement for making steel.

As a result, electric steelmaking increased its share of the market from about 12 percent in the late 1960's to 31 percent in 1982. Through the mid-1970s electric steel production complemented the growth of basic oxygen production. By the end of the period, electric steel production continued to increase its market share at the expense of both open hearth and basic oxygen processes. For the five-year period 1967–71, the average energy use intensity for making steel was 39.1 MBtu/ton shipped. For a comparable period ten years later, 1977–81, the average intensity of energy use had declined by almost 15 percent. As Figure 2-6 shows, by the end of the 1980s, electricity's share increased an additional six percentage points and average energy intensity had dropped another 15 percent.

This recent decline in the average energy use intensity seems modest until marginal differences are compared. For example, the amount of steel shipped in 1967 and 1980 was the same, 83.9 million tons. However, compared with 1967, fifteen million fewer tons of raw steel were produced in 1980 while 9.3 million more tons of scrap were used. (The remaining difference was made up by the improved production efficiency of electric furnaces and mills.) The energy implications of this shift are profound. To do the same job in 1980, the steel industry used 28.1 million tons less coal and the energy equivalent of another 1.9 million tons less coal in the form of oil and gas than in 1967. The extra electricity used was 13.6 billion kWh. This quantity of electricity can be generated from the equivalent of only 5.3 million tons of coal.

Thus, the same amount of steel was produced in 1980 as in 1967, but by using the equivalent of twenty-five million tons less coal. A steady continuation of this transition toward highly efficient production of steel is an essential part of any plan to regain and maintain the viability of the U.S. steel industry.

Nonferrous Metals Industries

Nonferrous metals industries include those that produce aluminum, copper, and cement.

Aluminum production

Aluminum and copper producers comprise the two largest energy consumers in this subgroup and both are steadily electrifying their processes. Because the industry now reduces alumina electrolytically, aluminum production is already largely electrified. In addition, processors now use electricity to replace fuels in the remelting of ingot and scrap. Processors prefer electric melting with both electric resistance heating elements and induction furnaces over combustion processes. Electric melting efficiently uses energy and reduces scrap losses to oxidation (up to 50 percent of the charged metal) because this process limits exposure to combustion gases.

Copper production

The copper industry can reduce the energy used in ore smelting by 40 percent when using an oxygen enrichment process; this process doubles electricity use. Electric smelting has replaced fuel smelting altogether in some installations. Even without electric processing, electricity will become increasingly important for copper production because U.S. ores are low grade and require either additional crushing and grinding or hydrometallurgical processing (i.e., acid leaching followed by electrolysis of the copper solution). Processors prefer this hydrometallurgical process when the ore concentration is below 0.3 percent: 10 percent of the U.S. copper industry employs this method. The fraction of energy purchased as electricity for the production of primary copper increased from 18 to 23 percent from 1974 to

1981. In the same period, the use of electricity per unit of production increased 1 percent while the direct use of fuel declined 24 percent.

Cement production

Increasing electrification of the cement production process presents some technical difficulties. Producers can replace gas with coal in the process without degrading cement (and some have begun to make this substitution). However, this switch to coal produces an associated (but modest) increase in electricity use, because conveying and pulverizing the coal requires electricity as well as do electrostatic precipitators that must clean up flue gases. Cement producers have begun using oxygen to enrich the combustion air, since it reduces total energy use by about 15 percent. If the oxygen is generated at the cement production site, the extra electric energy needed is about 10 percent of the fuel saved.

Plasma arc processing can produce a wide range of hydraulic cements and offers a promising use of electricity in this industry. Processors can incorporate plasma arc reactors into existing conventional cement kilns to provide additional thermal energy. Such a use is analogous to augmenting the output of fuel-fired glass furnaces with electric heat. Estimates show that plasma arc methods cost approximately 25 percent less than conventional methods.

Chemicals Manufacturing Industry

Electricity demand for chemicals manufacturing is driven both by electrification and by growth in output. Purchases of electricity by this industry increased 10 percent from 1974 through 1981; concurrently, the use of purchased fuels declined about 13 percent.[2-7] For example, the chemicals manufacturing industry can use electricity to produce oxygen for the partial oxidation of fuels. The advantages are exemplified by the "crude-oil cracking" technology to produce ethylene, introduced jointly by

Union Carbide Corporation and Kureka Chemical Industry, Ltd. In this new technology, electricity partially oxidizes deasphalted crude oil using steam at 2,000°C compared with 800°C in conventional steam crackers. The advantages of this high temperature conversion include increasing the yield of ethylene and coproducts from about 40 percent to about 65 percent and reducing residence time in the cracker from seventy-five milliseconds to seventeen milliseconds. Ethylene is one of the basic building blocks of plastics production, which represents the fastest growing sector of the industry.

Other applications include EXXON's oxo-alcohol plant in Baton Rouge, LA. as well as a joint venture between DuPont and U.S. Industrial Chemicals Companies in Deer Park, TX., which produces methanol, acetic acid, and vinyl acetate. The incentive for both projects is to use heavy residuum from petroleum refining to produce high value chemicals rather than to use more expensive natural gas. Elsewhere in the world, the partial oxidation of asphalt and residuum replaces high cost naphtha primarily in the production of hydrogen for ammonia synthesis.

The Hüls Company of Marl, Germany, SKF, and AVCO are now developing plasma arc technology to produce acetylene and synthesis gas directly from coal. In addition to the advantage of operating at very high temperature, electric energy shifts the carbon monoxide to hydrogen ratio in the synthesis gas. Estimates show that plasma production of acetylene from coal (as opposed to ethylene) can produce cheaper vinyl chloride.

Thus, because of its experiences with partial oxidation and plasma arc processing of coal and residuum, the chemical industry is adopting electricity-based methods to produce hydrogen (for ammonia and methanol) and the olefins—ethylene, propylene, and butadiene. Since these basic chemicals are used for synthesizing nearly half of the top fifty highest tonnage chemicals, the potential exists to extend continually the role electricity plays in the processing of organic chemicals. As chemical feedstocks are shifted from refined petroleum, natural gas, and natural gas liquids to coal and residuum, the oxygen needed for steam reforming nearly doubles. To the extent that these low-value feedstocks displace fuels and produce chemicals suitable for transportation fuels (e.g., methanol to increase the octane of unleaded gasoline), the industry can extend both the quantity of petroleum and its proficient use to serve the energy requirements for providing transportation services.

Medical Electronics (2-8)

Applications for electricity in medicine have offered some of the greatest advances for improving the quality of life for every human being. In every aspect of modern medicine, electrically powered equipment diagnoses and treats illnesses; virtually every modern medical tool substantially enhances the profession's capability to save and give life. Today, medical technology employs electricity in a myriad of ways. For example, doctors use it to stimulate muscle tissue and nerves to relieve pain, reinstate hearing in the deaf, and maintain a regular heart beat. In addition, a new and particularly exciting application involves using electricity to combat neuromuscular disorders that cause paralysis.

Researchers learn daily about the effects of small electric impulses on pain. Pain is an important indicator of where we hurt and what is wrong. Researchers are finding that electrical stimulation of muscle tissues and nerves causes the body to release naturally produced opiates—called endorphins—which either suppress pain or block the transmission of pain signals to the brain. This naturally-induced "pain killer" often relieves pain more efficiently than pharmaceutical drugs, many of which simply mask it.

Researchers have devised electrical systems that stimulate hearing in some profoundly deaf people. These systems consist of a microphone and transmitter located outside of the person, an implanted receiver, and an electrode located near the inner ear. The external system picks up sound, transmits it inward, and the receiver converts the signals for the inner ear to decipher.

Millions of people rely on pacemakers to regulate their heartbeats. This technology, one of the most commonly cited examples of electricity's ability to improve one's quality of life, has been in use since 1958. New generations of this technology can pace both the heart's right ventricle and atrium, and allow telemetry of the heart's activity.

Another recent, exciting application of electricity in medicine involves neuromuscular stimulation (NMS). NMS involves electrically stimulating a muscle so strongly that it contracts. Researchers are finding that this action often produces a powerful motor reaction which restores some level of function to dysfunctional limbs. Basically, NMS may enable para- and quadriplegics to walk! Thus, electricity may successfully treat several neuromuscular disorders; this research is rapidly evolving and early results prove quite promising.

Additional Applications Using Electrotechnologies (2-9)

Process applications to create ultraviolet light as well as ultrasonics, electrostatic charges, and magnetic effects require electricity. Typically these applications use small amounts of electricity to enhance, improve, or control conventional processes. Many of the applications address environmental problems because electricity use mitigates them in ways that do not add additional problems.

Ultraviolet (UV) light breaks molecular bonds to induce or catalyze chemical reactions. The bond-breaking can kill unwanted organisms during wastewater treatment. Ultraviolet light also excites chemical oxidants, such as ozone, thus increasing their power to oxidize. These UV techniques operate at room temperature, reduce residence time, and have low capital and operating costs. The technique effectively destructs dioxin, polychlorinated biphenyls (PCBs), and other chlorinated hydrocarbons as well as organisms and viruses in wastewater.

Industry now uses ultrasonic wave generators in liquid spray applications. The sonic-atomized fluid has several advantages compared with pressure spray systems. The sonic sprays at low velocity, which virtually eliminates overspray and related material losses. The spray also produces very small droplets that are more efficient in applications such as spray-drying and humidifying. The sonic atomizing technique does not force liquid through a small aperture, so the technique can handle a variety of liquids and slurries at widely varying flow rates without clogging. For example, sonic water sprays can increase lime kiln throughput by 10 percent. With ultrasonic spraying, the fine water mist evaporates before the lime and water can react. Thus, the technique cools kiln flue gas without clogging the flue gas filter bags. This action, in turn, permits kiln operation at a higher temperature than would otherwise be possible.

Electrostatically enhanced scrubbing systems can clean up acid, oil, and resin mists; titanium dioxide dusts; and ammonium chloride, phosphorus pentoxide, and potassium chloride from off-gases. Charging aerosols or scrubber water mist electrostatically improves their ability to capture and remove submicron-sized particles. The technique improves scrubber efficiency from 50 percent to 96 to 98 percent and costs only a fraction of equivalent nonelectric scrubbing systems. Industry is applying the same general principle to enhance particle removal efficiency from viscous liquids.

Electrochemical synthesis

Chemical producers spend a majority of their capital and operating budgets on the reactors that convert coal or oil into organic chemicals. These city block-sized, three-dimensional circulatory systems contain an array of piping and connected equipment that separates, recycles, and recovers chemicals. This huge system (both in size and cost) can impede a manufacturer and often provides the incentive to develop electro-organic chemical synthesis systems that offer the potential to yield pure specific chemicals.

Today, the cost of electricity restricts the use of electro-organic synthesis. Only producers of small volume, high value chemicals with high molecular weights can profit using this electrotechnology. Currently, some twenty electro-organic chemicals are produced commercially and ten more are in various stages of development. Much of this technology's development occurs outside the U.S. because many countries' (e.g., India) producers work on a much smaller operating scale than more developed countries.

Despite electroprocessing's relatively high costs, it confers advantages in product quality and process cleanliness that are not possible with thermal techniques. For example, the conventional halogenation route to anisic alcohol leaves trace contaminants with unpleasant odors—a problem if the alcohol is used to impart fragrance. This problem does not arise from use of the electrochemical method. Electrosynthesis of isocyanates and carbamates eliminates the use of highly toxic phosgene that would be needed in thermal processes. This "process cleanliness" is also a major advantage of using electro-organic processing to produce Vitamin C—in this case, manufacturers do not have to use a hypochlorite oxidizing agent, which is troublesome to dispose of.

An electrochemical cell can gasify coal; the anodes release oxides of carbon and the cathode releases hydrogen. Since coal provides about half of the energy for the gasification, use of this electroprocessing method is similar to using plasma arc technology.

Recycling

The cost of materials, including chemicals, fuels, and electricity, dominates production costs throughout manufacturing. Thus, the chemical industry and its related subsidiaries are developing and adopting ways to reduce these costs and material recovery and recycling offers one viable option. Since material recovery and recycling is often performed on small scale in a decentralized fashion, electric processing often offers a valuable method for accomplishing it.

New Applications for Mechanical Energy

Operations accomplished by physical as opposed to thermal or chemical processes typically use less fuel to provide the mechanical energy. Rapidly rising fuel costs make new applications for mechanical processes more attractive. Further, since utility-generated electricity uses low value fuels such as coal and uranium, new applications for mechanical processes are energized to an increasing degree with utility-produced electricity.

New mechanical process applications are enormously varied and ubiquitous; however, not all potential applications for mechanical processes are equally attractive. The main objective of adopting these process applications is to reduce the overall cost of production. A particular application, such as using a heat pump in lieu of a gas burner and heat exchanger, may save energy dollars but cost more to purchase and maintain, and thus fail to satisfy justification criteria. Another application, such as the use of membrane separators in lieu of chemical absorption, may reduce each of the production cost components (i.e., capital, labor, and energy) and also provide other benefits such as improving product yield or purity.

Examples of beneficial mechanical processes include installing:
- Vacuum pumps to replace steam ejectors
- Rotating stages in distillation columns to reduce the numbers of stages required
- Freeze crystallization in water processing to reduce energy costs for evaporation
- Ultrafiltration to improve a wide variety of waste water clean-up chores
- Pressure swing absorption and cryogenic processes to recover hydrogen and argon from purge gas in the production of ammonia
- Physical absorption of carbon dioxide in propylene carbonate to eliminate steam stripping of potassium carbonate solution in the production of ammonia
- Pressure swing absorption to displace wet process technol-

ogy in the separation of pure hydrogen produced by steam reforming

Using semipermeable membrane separators offers an important application of mechanical processing. These separators are packaged in the form of a shell and tube bundle of thousands of hollow polysulfane fibers. They substantially improve the efficiency of separating gas mixtures. Current applications include recovering hydrogen from purge gas during production of ammonia, separation of acid gas (carbon monoxide and hydrogen sulfide) and carbon dioxide (for tertiary oil recovery) from natural gas, and enrichment of air in nitrogen or oxygen. The possibility of impregnating the semipermeable membranes with specific materials greatly expands the potential applications for the technology. The Amoco Research Center has obtained good separation of ethylene or propylene from methane streams with cellulose acetate fibers impregnated with silver halide. A joint venture between Solvay and CIE (Brussels) and SNIA Fibre (Milan) is developing membranes containing enzyme-loaded organisms. This concept allows for simultaneous product reaction and separation.

References:

2-1 *Electricity and Industrial Productivity*, Electric Power Research Institute EM-3640, 1984.

2-2 "Freeze Concentration: An Energy Efficient Separation Process," *EPRI Journal*, January/February 1989.

2-3 "Electric Van and Gasoline Van Emissions: A Comparison," EPRI Technical Brief, TBCU-177, October 1989.

2-4 "Telecommuting: A National Option for Conserving Oil," CRS Report for Congress, Congressional Research Service, The Library of Congress, Report 90-524SPR, November 9, 1990.

2-5 *Roles of Electricity: Electric Steelmaking*, Electric Power Research Institute, EU-3007-8-86, 1986.

2-6 *Roles of Electricity: Recent Trends in Manufacturing*, Electric Power Research Institute, EU-3013-3-87, 1987.

2-7 *Electricity in the American Economy: Agent of Technological Progress*, S.H. Schurr et. al., page 153, Greenwood Press: New York, 1990.

2-8 "Medical electronics improve quality of life for patients," *Design News*, February 3, 1986.

2-9 *Roles of Electricity: Production of Chemicals*, Electric Power Research Institute, EU-3015-7-86, 1986.

SECTION II.

Electricity's Environmental, Political, Social, and Economic Impacts

CHAPTER 3
Environmental Impacts and New High Value Solutions

Electricity yields many rewards but, as with all energy production, it significantly impacts the environment. In particular, electricity production and use plays a role in several of today's major environmental concerns, including the "greenhouse effect," global warming, and acid rain. Electric utilities must directly address these issues, because ignoring them only diminishes customers' perceptions of electricity and its value.

This chapter briefly discusses several current environmental concerns. Then, it offers some examples of electrically powered technologies that reduce the environmental impacts of space conditioning as well as several industrial processes. As these examples (and additional ones presented in other chapters) show, efficient uses of electric technologies can reduce the environmental impacts of many processes, which only increases the actual and perceived value of electricity.

Environmental Impacts of Energy Production

The term "greenhouse effect" refers to the increasing concentration of heat-trapping gases in the earth's atmosphere. The most important of these gases, carbon dioxide (CO_2), is responsible for about half of the effect. In addition, methane (CH_4), nitrous oxide (N_2O), and CFCs combine with CO_2 to act as a "shield" that traps heat within the earth's atmosphere (Figure 3-1). Many scientists believe this growing concentration of greenhouse gases in the atmosphere continually increases as a result of human activities, although normal biological cycles naturally produce the gases also. The greenhouse effect has clearly emerged as an important public policy issue in the 1990s. (3-1)

Although many uncertainties remain about the greenhouse effect and its impacts on the earth's climate now and in the future, scientists largely agree on the concept and its ultimate environmental consequences. All major climate models predict that the increasing level of atmospheric greenhouse gases (equivalent in heat-trapping capability to a doubling of the preindustrial level of CO_2) will cause the earth's average temperature to rise. Estimates for this "global warming" range from 1.5°C to a high of 4.5°C because of uncertainties about differences in the earth, air, and ocean cycles. These conditions, depending on their level and degree of change, could exacerbate or dampen global warming, thus causing the temperature changes to deviate from this current range in predictions.

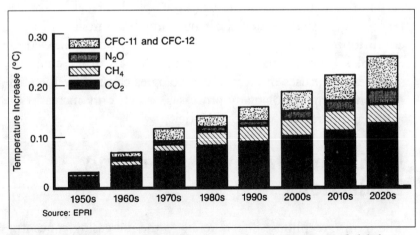

Figure 3-1. Historical and projected increases in greenhouse gases and global temperature levels.

The low end of the computer model estimates—a 1.5°C warming—would create the warmest conditions experienced in recorded history. A 4.5°C rise would create a temperature level last experienced in the Mesozoic era—the age of dinosaurs. One thing is clear: once the gases are emitted to the atmosphere, the effects may be irreversible.

Scientists expect that the level of atmospheric CO_2 will effectively double within fifty to 100 years. Concentrations above a CO_2 doubling could yield even greater warming. Some researchers believe that greenhouse gas emissions to date have already warmed the earth from 0.5° to 1.5°C. Recent analyses of temperature records over the past century reveal an overall warming of about 0.5°C, with the three hottest years occurring during the 1980s and the nine warmest since 1978.

More than half of greenhouse emissions comes from the extraction and use of fossil fuels in vehicles, buildings, factories, and power plants. Fossil fuel emissions also contribute to acid rain, urban smog, and stratospheric ozone depletion. Many regard electric utilities, via electricity production, as major contributors of CO_2, sulfur oxides (SO_X), and N_2O (which is implicated in the breakdown of ozone). The release of CO_2 and nitrogen oxides (NO_X) into the atmosphere from power plant stacks has been identified as a primary cause of acid rain. In addition, vehicle emissions such as volatile organic compounds (VOCs) and NO_X—both precursors of ozone—and carbon monoxide (CO) contribute principally to urban air quality problems.

Despite the remaining uncertainties about the extent and impacts of the greenhouse effect and global warming, interested parties have articulated many major policy options for combating these emissions. Options focus on slowing the build-up of human-made heat-trapping gases in the atmosphere and eliminating CFCs, phasing out HCFCs, and limiting fossil fuel combustion. Many nations have agreed to phase out the production of CFCs and certain other ozone-depleting chemicals.

This agreement, called the London Amendment to the Montreal Protocol, asks industrial nations to stop producing CFCs by the year 2000 and HCFCs by 2030. The U.S. is seeking to meet the CFC reduction goal by the mid–1990s.

In addition, the 1990 Clean Air Act will impact emissions. Amendments to the Act set up an emission trading allowance for sulfur dioxide (SO_2). This trading system will also have important implications for utility DSM efforts: utilities will be able to meet emissions requirements through reductions or by purchasing emission allowances from other companies. In addition, DSM programs that promote the use of energy efficient electrotechnologies will help utilities and their customers comply with these new environmental regulations.

In October of 1992, the U.S. congress passed the National Energy Policy Act, which mandates nationwide energy efficiency standards and restructures the way utilities are regulated in this country. This act strengthens existing efficiency standards for appliances and automobiles, among other things. It recognizes that while the focus to date has been on supply-related issues, experts must also incorporate demand-related issues—efficiency standards—in today's analyses to fully assess environmental impacts and opportunities. *Thus, the core of the debate about solving the greenhouse effect and global warming centers on issues of energy policy.*

Under current plans to control acid rain, many existing and new power plants will be required to use SO_2 and NO_X reduction technologies. By the year 2000, use of such technologies could decrease annual power plant SO_2 and NO_X emissions by nine million tons and two million tons, respectively. Implementing emission reduction strategies would, however, increase the cost of electricity in some regions of the U.S. Policymakers have also examined the possibility of controlling the biological sources of greenhouse gases, however the practicality of such measures remains highly uncertain.

Many of these most pressing environmental challenges respect no political boundaries. Global warming, stratospheric ozone depletion, and acid rain are *global* problems that individual countries can solve only with unprecedented international cooperation. Similarly, hazardous waste management, urban air pollution, and fouled waterways are regional problems that are so pervasive they demand national attention to achieve significant environmental improvements. Electricity offers many important opportunities for reducing or eliminating the volume of pollutants entering the biosphere. One crucial step: begin conserving resources and improving the efficiency of their use via new technologies.

Consensus on achieving these dual goals may be difficult to reach. However, a diverse group of individuals has proposed specific measures that include:

- Implementing carbon taxes (to discourage use of high CO_2-releasing fuels like coal and coal-based synfuels)
- Creating stricter vehicle mileage per gallon of fuel standards
- Requiring environmental impact statements to include a discussion of the climate implications of energy projects.
- Encouraging fuel switches to natural gas and/or nuclear power
- Using renewable energy sources such as wind, solar, and annually-cycled biomass
- Mandating wide-scale electric utility energy efficiency programs

Some believe that an ambitious, cost-effective program of efficiency improvements in the residential, commercial, industrial, and transportation sectors could keep world energy consumption essentially flat for the next thirty years. This estimate even allows for economic growth and population increases!

Existing Electrically Powered Technologies Meet Stiff Environmental Challenges

Many electrically powered technological advancements reduce the environmental impacts of older, outdated systems. Consumers (both residential and business) that adopt these new electric technologies recognize the value of electricity—both in improving energy efficiency and limiting environmental damage. This section discusses several of these technologies. The chapters in Section III discuss industry-specific advancements in greater detail.

The Electric Heat Pump

The electric heat pump is one of the single greatest electrically powered technological advancements. Since the early 1970s, electric heat pumps have doubled their cooling efficiency and increased heating efficiency by 50 percent. Electric heat pumps are available today with a steady-state efficiency almost four times that of an advanced gas-fired condensing furnace. One such residential heat pump now on the market is Carrier's "HydroTech 2000," the result of a six year multi-million dollar research program sponsored jointly by EPRI and Carrier.

Today's heat pump technology evolves continually and several features make the heat pump the most efficient space conditioning and water heating system available today. These innovations include electronic variable speed compressors and blower drives, a novel coil defrosting method, a patented refrigerant circuit for year-round integrated water heating, and, above all, the introduction of rapidly evolving electronic drive and control technology. These advances bring the highest ever comfort and energy savings to utility customers as well as offer tremendous environmental benefits to society.

Even more dramatic heat pump technology improvements are

almost assured. The 1987 National Appliance Energy Conservation Act established minimum heating and cooling efficiency standards for heat pumps, which take effect in 1992 (split-systems) and 1993 (single-package equipment). Because over 90 percent of the heat pumps currently being manufactured and marketed fall below the federal minimum standard, significant increases in heat pump efficiency must occur in the very near future.

Although the heat pump's performance during the market's infancy was anything but smooth, three decades later, designers have corrected most of the problems. The electric heat pump has established itself as a reliable, energy efficient and cost-effective heating and cooling option, able to meet the tests of manufacturability, consumer acceptance, and reduced environmental impacts.

Twenty-six percent of the new single family homes built in 1988 were equipped with an electric heat pump; in new multifamily dwellings, the figure was even higher. Heat pump shipments in the U.S., while somewhat dependent on economic cycles—especially new housing starts—have, nevertheless, been rising steadily. Annual air-source electric heat pump shipments in recent years have risen to the 900,000 unit level; several hundred thousand more water-source, packaged terminal, and other types of heat pumps are shipped annually than in the 1970s.

Electric heat pumps versus gas systems: one comparison

Efficient electric heating and cooling systems use less resource energy and emit less CO_2 than high efficiency gas heating systems coupled with electric cooling. The reason: the electric heat pump is the most efficient integrated heating/cooling technology available today. It is more efficient to use natural gas to generate the electricity to operate a heat pump than to burn the gas directly for heating and use electricity for air conditioning.

EPRI recently conducted a study of the resource energy requirements of the electric heat pump compared with those of a gas/electric and all-gas system. This study tested each system's efficiency when heating and cooling an 1800 sq. ft. home in Philadelphia (representative of the Department of Energy's [DOE] Region IV) with an annual heating requirement of forty-six MMBtu and an annual cooling requirement of nineteen MMBtu. (DOE Region IV represents a moderate-to-cold, heating-dominated climate, which acts as the basis for the U. S. government's rating of heat pump performance. The study assumed that the source of electricity supply was an efficient gas-fired combined cycle power plant (heat rate 8140 Btu/kWh). Pipeline, transmission and distribution losses were taken appropriately in all cases.

First, the study compared the electric heat pump with one of the most efficient furnaces available—the pulse combustion condensing furnace. This furnace offers annual fuel utilization efficiencies (AFUEs) ranging from 92 to 96 percent, depending on rated capacity. Auxiliary electric power requirements for the air circulating blower, purge blower, spark plug igniter, and controls run from about 400–700 watts. For the purpose of this study, researchers assumed an AFUE of 95 percent, with auxiliary electrical requirements of 400 watts. For air conditioning purposes, they coupled a high efficiency electric air conditioner, with a seasonal energy efficiency ratio (SEER) of 13.5, with the gas furnace.

Researchers used the Carrier "HydroTech 2000" electric heat pump in this study. In Philadelphia's climate, its annual electricity consumption was estimated at 5,146 kWh for heating and 1,291 kWh for cooling. When the study was completed, researchers discovered that the gas furnace/electric air conditioner combination consumed 19 percent more resource energy (gas) than the electric heat pump. (3-2)

To meet the heating requirement of the house, the pulse combustion furnace required 48.5 MMBtu of gas and 335 kWh of electricity for the auxiliaries; the air conditioner consumed 1,396 kWh of electricity. Converting the electric energy to the gas fuel consumed by the combined cycle power plant yields the comparison illustrated in Table 3-1.

Table 3-1
Electric Heat Pump vs. Pulse Combustion Furnace/Electric Air Conditioner—Resource Energy Requirements in MMBtu

	Electric Heat Pump	Gas Furnace Electric Air Conditioner
Heating, Electricity	41.9	2.7
Heating, Gas		48.5
Cooling, Electricity	10.5	11.4
Total	52.4	62.6

Second, the same high efficiency electric heat pump was compared with a hypothetical gas heat pump, whose performance characteristics were similar to the Gas Research Institute's (GRI) near-term market entry unit. Researchers assumed this heat pump had a seasonal average heating coefficient of performance (COP) of 1.4, a cooling season COP of 0.8 and auxiliary power requirements of 1,000 watts (for air circulation blower, radiator fan, jacket cooling water pump, starter motor, and other auxiliary equipment). These auxiliary requirement performance levels have reportedly have been achieved in the laboratory.

As before, researchers converted the gas heat pump's auxiliary electrical consumption, 658 kWh in the winter and 512 kWh in the summer, to resource energy, assuming a heat rate of 8,140 Btu/kWh. Table 3-2 illustrates that the gas heat pump consumed 27 percent more resource energy (gas) than the electric heat pump.

Table 3-2
Electric Heat Pump vs. Gas Heat Pump—Resource
Energy Requirements in MMBtu

	Electric Heat Pump	Gas Heat Pump
Heating, Electricity	41.9	32.9
Heating, Gas		5.4
Cooling, Electricity	10.5	4.2
Cooling, Gas		23.6
Total	52.4	66.1

This study verifies that the gas heat pump is indeed an efficient heating device. However, its poor cooling performance and high auxiliary electrical power requirements make it less efficient in overall resource terms than an efficient electric heat pump. Moreover, even a high efficiency gas furnace/electric cooling combination is more energy efficient than the gas heat pump. It is obvious from Tables 3-1 and 3-2 that the electric option emits less CO_2, since its total gas resource consumption is lower than either the gas/electric or the gas heat pump option.

An ongoing EPRI study has calculated CO_2 emissions for a range of heat pump efficiency cases and for several assumptions about electric generation, including existing coal-steam, current average U.S. utility fuel mix, combined cycle (similar to the case analyzed above), and the fuel cell. Figure 3-2 summarizes these results, and plots the ratio of the CO_2 emissions of the gas end-use option to the electric option as a function of electric heat pump efficiency (taking the SEER as the proxy for the latter). This figure illustrates conditions in DOE Region 3. It compares a gas system that includes the high efficiency pulse combustion gas furnace and a high efficiency air conditioner with the electric heat pump. (3-3)

The emissions ratio of unity divides the figure into two horizon-

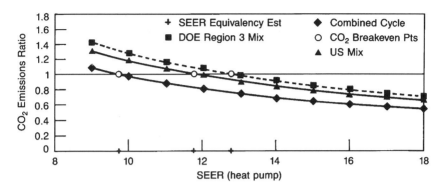

Figure 3-2. Influence of generation mix and type on CO_2 emission ratios of electric versus natural gas space conditioning.

tal areas: when the gas-to-electric emissions ratio is greater than one, the gas option emits less CO_2; a ratio of less than one indicates a preference for the electric heat pump, due to lower CO_2 emissions. The figure shows that a high efficiency electric heat pump results in lower CO_2 emissions than the high efficiency gas/electric system. This finding holds whether the electric energy source is coal, the current U.S. average fuel mix (24 percent of which is nuclear and hydro), or combined cycle.

Industrial Electrotechnologies

Many U.S. industries discharge toxic pollinates into U.S. rivers, lakes, oceans, and waterways. Companies in the industrial sector emit approximately one-third of the CO_2 and nearly all of the SO_2 and CFCs in the nation. Since, overall, the industrial sector dumps the greatest percentage of pollution into the environment, it represents the greatest opportunities for cleaning up the environment.

In 1990, the industrial sector consumed 35 percent of U.S. electricity; the steel, aluminum, chemical, glass, petroleum, food, textile, and paper industries used most of it. Motors used in industrial applications consumed nearly 70 percent of this total. The remaining 30 percent includes electrolysis, lighting, process heating, and space heating applications.

While many of the newer industrial electrotechnologies were developed before environmental awareness heightened, they still meet today's tougher standards. The very features that make electrotechnologies more efficient and economically attractive also frequently make them more environmentally benign. Thus many electrotechnologies offer multiple advantages in most industrial applications; they often minimize pollution, enhance productivity, and offer greater economic benefits.(3-4)

Electrotechnology research and development projects currently focus on developing technologies that fit into three categories: energy efficiency, waste prevention, and waste treatment. Energy efficiency reduces global energy needs and the resulting pollution caused by excess generation and inefficient equipment. Waste prevention technologies stop pollution "before the fact." Finally, waste treatment technology cleans up the waste after it has been generated—"after the fact."

For example, electrostatically applied powder coatings with infrared curing *replace* solvent-based painting altogether. This technology eliminates solvent effluent "before the fact"—before it is generated. On the other hand, a solvent recovery system takes solvent-laden air and reduces its hazardous emission levels, and allows reuse of the recovered solvent. This system cleans up a problem—"after the fact." Research and development must progress in each of these categories if industry is to effectively and economically eliminate and/or minimize hazardous waste discharge.

Industrial applications of electrotechnologies can be grouped into four areas: **materials production, materials fabrication, process industries, and waste and water treatment**. These areas are discussed in further detail below.

Materials production

This industry category offers an excellent example of how electrotechnologies can produce an "after the fact" solution to a hazardous waste problem. Electric arc furnaces (EAFs) now account for about 40 percent of total steel production and their use will continue to grow steadily into the next century. However, EAF steelmaking generates about twenty to forty pounds of dust per ton of steel. EAF dust particles are primarily composed of iron and zinc oxide as well as small amounts of leachable lead, cadmium, and chromium. Because they contain these ingredients, the U.S. Environmental Protection Agency (EPA) has classified these dust particles as a hazardous waste. Traditionally, steelmakers have placed EAF dust in landfills, but the 1984 amendment to the Resource Conservation and Recovery Act prohibits such disposal. Accordingly, pollution control is a major issue facing foundries today.

The coke-fired cupolas used by many foundries create a number of environmental problems, including atmospheric pollution and the need to dispose of hazardous waste. To mitigate air pollution, coke-fired cupolas require expensive pollution control equipment; the waste disposal problem, however, remains increasingly insoluble. Steelmakers do have options—one in particular offered by an electrotechnology.

A new Tetronics plasma-based process recovers zinc, lead, and cadmium while producing a non hazardous, readily disposable, 100 percent benign slag. Additionally, this process practically eliminates air pollution problems at the foundry site because these systems do not need combustion air or gas cleaning devices, such as scrubbers. These electric induction melting furnaces are highly efficient, precisely control the rate of heat input, and provide a quieter and cleaner plant environment—when combined, these factors make them extremely competitive with coke-fired cupolas.

Large scale pilot tests have shown that the plasma-fired cupola offers major advantages over the conventional coke-fired cupolas for scrap iron remelting. In a plasma-fired cupola, plasma torches provide heat for melting which substantially decreases the requirement for coke and other fuels and lessens the combustion air requirement. This lower combustion air requirement greatly diminishes pollution problems at the foundry site. In addition, the lower air velocities and volume required by the system allows it to melt relatively inexpensive iron borings and steel turnings without blowing them out of the furnace. Moreover, the melt chemistry is easier to control without the heavily oxidizing airflow of conventional cupolas. These factors result in reduced operating costs and improved air quality. General Motors has installed and now operates the first commercial installation of this technology at its Central Foundry Division in Defiance, Ohio.

Other foundry applications of electrotechnologies include reclaiming sand used for casting and recovering/removing chromium from waste streams. Sand used in the casting process is "bound" (held together) with "binders" of clay-like substances. Reclaiming this sand for future use is currently of interest since the costs of purchasing new sand and disposing of used sand continues to rise. (In addition, increasing regulations for solid waste disposal also poses problems for foundries.) Gas-fire sand reclamation units are in limited use in the domestic foundry industry. Electric-fired units, which are relatively new, have been used in Canada and Europe. Electric-fired units potentially offer lower capital cost, more flexibility, less maintenance, and smaller size unit design options.

A recent application of an electrotechnology includes a project to examine processes for recovering/removing chromium from waste streams. In this process, chromium plating wastewater, containing hexavalent chromium, flows into activated carbon columns where the chromium ions collect on the carbon. The clean solution can either be dumped down the drain or recirculated to the plating operation. After a period of time, a dilute solution of sulfuric acid is pumped through the columns stripping off the chromium, leaving the carbon clean, and forming a solution rich in trivalent chromium. This solution is pumped into a holding tank and subsequently is processed through electrowinning cells to plate out chromium oxides on the cathodes. The chromium oxide can then be sold to a processor and the proceeds used to pay for the operation.

Materials fabrication

Electrotechnology applications in materials fabrication range from the use of the infrared and ultraviolet spectrums to applications of induction. For example, the auto and printing/paper industries fabricate materials. Traditional solvent-based methods of curing, drying, coating, and painting release large volumes of VOCs into the environment. Now, the advent of precise infrared and ultraviolet electrotechnologies eliminates this environmental hazard.

For example, the Chrysler Corporation recently tested electric infrared (IR) auto coating and drying. This technology uses computer controlled high intensity IR drying to flash-off the solvents/water from base coat paints. Early results show this technology can highly improve surface finish and will eventually allow the use of water-based paints, which will replace traditional solvent-based paints. This substitution reduces organic solvent emissions and meets environmental compliance standards. Chrysler's Belvedere Assembly Plant uses IR to produce new mirror-like finishes on its Chrysler New Yorker, New York Landau, and Dodge Dynasty lines. Based on its experiences, Chrysler intends to convert from a solvent-based basecoat to a water-based solution. The IR drying process also drives out 95 percent of the water prior to applying the clearcoat.

Another IR application involves replacing traditional liquid paints (many of which are solvent-based) with "powder coating." Technicians apply powder coating by spraying electrostatically-charged powder onto a surface; electric IR heating then cures the surface. Since technicians apply this coating electrostatically, they can easily control thickness and uniformity. In addition, powder overspray is trapped and recycled; eventually 95–99 percent of the powder is actually applied to the surface so very little clean-up is required. In the old method only 35–40 percent of the liquid paint actually adhered to a surface. The other 60–65 percent was wasted and technicians constantly had to scrape and scrub the paint area. Also, because the powder coats only the product and contains no solvents, it emits no toxic fumes or leaves a solid waste residue to remove; accordingly, the process is environmentally benign.

The Coors Brewing Company now cures its can coatings using ultraviolet radiation (UV). A conventional printer applies wet UV inks to the can and then adds a clear UV varnish as the final layer. The inks and the varnish cure in about half a second when exposed to UV radiation, as opposed to the twelve seconds required by traditional gas convection oven curing process. By adopting this UV process, Coors eliminated the need for 650 cubic feet per hour of gas consumed by the conventional oven and replaced it with only moderate electricity consumption. Additionally, the conventional gas oven's exhaust gases required incineration to minimize emissions, whereas the UV system emits no fumes and poses virtually no environmental threat.

Process industries

Various electrotechnologies now help process industries treat waste streams as well as increase the energy efficiency of several processes (both applications enhances electricity's value). The potential for expanding electricity's use in these industries grows daily. For example, electricity is underrepresented in process heating. Currently, electricity provides only a fraction of one percent of the 2,000 trillion Btu used to heat, cook, soften, melt, distill, cure, and fuse materials and products in the process industries.

One growing area is the use of heat pumps to recover solvents. For example, the Brayton cycle heat pump (BCHP) recovers and recycles VOCs and solvents. Environmental legislation and economics require control of the emissions of VOCs, including CFCs, from process waste streams. The control of VOC emissions and the manufacture of chemical solvents for industrial processes accounts for a sizable amount of U.S. energy consumption. Existing technologies to control VOCs include thermal oxidation, catalytic oxidation, adsorption, and condensation by refrigeration cycle. The BCHP condenses VOCs to liquids by using a reverse Brayton refrigeration cycle. The BCHP cools VOC laden air streams to very low temperatures, then condenses and collects the VOCs. If recovered properly, the BCHP will not degrade the solvents, which can then be directly recycled or reused.

The 3M Company installed a first generation BCHP system at its manufacturing plant in Hutchinson, Minnesota. This application is considered a large system (greater than 10,000 CFM). However, most industrial needs for VOC recovery will probably have smaller BCHP size requirements. Current research is directed towards developing smaller sized (around 2,000 CFM), lower cost equipment. For example, a small, mobile BCHP is presently being fabricated for use at a variety of host sites. The initial target area for these test units is the Los Angeles area, where the implementation of ever stricter air and water pollution regulations demand solutions such as this small BCHP unit.

Industrial process heat pumps reclaim wasted energy and apply it in a cost-effective manner. Three basic types of industrial process heat pumps are currently used: the closed cycle, open cycle, and the semi-open cycle. The best type for a given application depends on process-specific factors.

The closed cycle heat pump system is appropriate for industries with relatively low temperature heating requirements. Typical applications include food processing, water purification, pharmaceutical production, plastic molding, glass fabrication, lumber drying, and wood preservation. The open cycle system uses steam or another process vapor as the working fluid, rather than utilizing a separate refrigerant. These systems tend to be larger than closed cycle systems.

Semi-open cycle systems, also known as mechanical vapor compression systems, are usually used to recover heat from contaminated waste streams. Currently, food, chemical, petroleum, pulp, and paper industries use semi-open cycle systems.

The benefits of a well designed industrial process heat pump system have been demonstrated in nearly 1,400 U.S. applications. The most obvious benefit is the lower energy costs which result from using an internally available source to meet primary energy requirements. Depending upon the specific application, a well designed industrial heat pump system can pay for itself in two years or less. While this potential for energy savings alone may warrant careful consideration of an industrial process heat pump system, further benefits include: lower temperature operations, greater process control, and improved equipment performance, environmental conservation, safety, and application of alternative technologies.

Electrotechnologies also offer the process industry several opportunities to increase energy efficiency. Recent advances in freeze concentration have helped the dairy industry, which is the largest food industry user of energy, markedly reduce its energy consumption. The equipment used now—most typically evaporators—is generally antiquated and not nearly as efficient as freeze concentration. In addition, it uses large supplies of fossil fuel as an energy source.

Applying freeze concentration to replace thermal evaporation yields superior quality dairy products; in addition, based on only 10 percent market penetration of this efficient technology, it could save 3.4×10^{12} Btu/year of fossil fuel. Obviously, reductions of this magnitude significantly lower pollution levels and foreign oil dependence. In addition, freeze concentration technology requires lower operating temperatures which reduce microbiological and enzymatic activity. Food quality improves, equipment is utilized more efficiently, and cleaning costs drop. Also, freeze concentration allows for longer operating cycles which require less frequent heat exchanger cleaning operations. Curtailing cleaning operations reduces wastewater generation, lowers waste treatment costs, and cuts the amount of milk solids lost with the wastewater.

Another dairy application, cheese production, results in about forty-two billion pounds of whey as a by-product. Whey contains over half the solids of milk, and its use as a food ingredient has steadily increased as a replacement for nonfat milk solids in the U.S. food manufacturing industry. Additionally, the solids in whey can be fermented to provide fuel grade alcohol. But despite significant gains in its use, half of the whey produced is not utilized and is still disposed of as raw waste. However, with freeze concentration technology, the solids contents of whey increases from 12 percent to 42 percent which raises its value from $.10/lb to $3.00/lb. This improved waste product becomes more valuable as a resource, recyclers use it more, and the amount disposed lessens.

The process industries yield numerous additional energy efficiency opportunities. Many of these opportunities appear via the use of an energy analysis tool called "pinch technology." Because industrial processes contain both heat sources and heat sinks, matching these effects through heat exchangers represents a major opportunity to save energy and improve efficiency. Pinch technology allows heating system designers to match heat sources with sinks; it tracks a process system's heat flow for an entire plant or a unit operation. Implementing pinch technology recommendations results in energy savings of approximately 25–40 percent and reduces energy consumption (including fuel burned), thus lowering emission levels and environmental pollution. In addition, since pinch technology lessens a facility's steam load, it also diminishes the amount of high temperature wastewater discharged into rivers, lakes and oceans. This high temperature wastewater reduction reduces thermal pollution.

Waste and water treatment

Electrotechnologies have yielded a number of innovative waste and water treatment applications. Stricter environmental regulations and the decrease of available space for waste disposal have motivated an ongoing reevaluation of waste management technologies and an emerging interest in innovative waste processing methods. Waste types can be divided into five major classifications: 1) municipal solid, 2) municipal hesitater, 3) non-hazardous industrial, 4) hazardous industrial, and 5) nuclear. Within each category, different types and quantities of waste use different levels of energy and require different disposal and/or treatment technologies.

Three areas offer a high potential to benefit from electrotechnologies: hazardous-waste treatment, hesitater treatment, and resource recovery. Researchers are studying several processes for hazardous waste management: pyrolysis (including plasmas and infrared heaters), electrochemical concentration (including electrodialysis), freeze concentration, and supercritical fluid oxidation. Processes examined for water treatment include ion exchange, reverse osmosis, and ultrafiltration. Plasma, reverse osmosis, and electrodialysis can be employed in metals recovery and for melting both glass and metal.

The treatment of liquid waste poses several major problems: it is difficult to handle, environmental protection regulations continually tighten, the available space for waste disposal decreases daily, and the amount of liquid waste generated increases steadily. The new environmental regulations emphasize source reduction, recycling, and treatment as well as impose tighter hazardous waste handling and disposal requirements. These regulations result in significant increases in disposal costs.

These combined factors have led to the widespread study, development, and use of concentration and separation techniques. These studies focus on industries that generate the largest wastewater streams and industries under the most regulatory pressure from federal, state, and local agencies. These industrial sectors include: food and kindred products (SIC 20), paper and allied products (SIC 26), chemicals and allied products (SIC 28), petroleum refining (SIC 29), and metal finishing (SIC 347), which includes printed circuit board manufacturing. Four of these five largest users of water (food, paper, chemicals, and petroleum refining) are in the *process industries category* and account for over 60 percent of the industrial water intake in the U. S.

The studies have also examined each sector and identified the capital cost, energy use, area of application, and equipment manufacturers associated with the liquid waste streams. In addition, researchers gathered information about the different manufacturing processes and the type and amount of resulting waste streams, and current industry practices for processing these wastes. The study identifies several existing and emerging separation and concentration electrotechnologies for treatment of industrial liquid wastes. Five technologies have emerged as the most promising:

1. *Crossflow microfiltration using ceramic membranes* effectively separates oil and water and is nearing commercial acceptance by industry. However, researchers still need to sponsor test programs to obtain membrane fluxes, separation factors, and run lengths. These findings will prove the membrane can separate the two elements without plugging.

2. *Supercritical water oxidation* may represent the best opportunity to almost completely destruct hazardous chemicals contained in an aqueous medium. This process operates at high pressure and temperature and requires expensive construction materials. Researchers need to sponsor test programs to demonstrate the process will run at severe conditions of pressure and temperature without excessive downtime. This technology will be expensive to commercialize because of relatively high operating and capital costs of even small scale units.

3. *Microwave distillation* can make distillation separations quickly for applications such as distilling pharmaceuticals and other thermally sensitive materials. Researchers need to identify cost-effective applications and obtain performance data.

4. *Electrically-driven drop dispersion* uses an electric field to break up and disperse water droplets in organic media; it could be applied to both liquid–liquid and supercritical extraction. The Oak Ridge National Laboratory is developing this technology. To speed commercialization, it must construct a liquid-liquid extraction device capable of processing five to ten gallons per minute, then obtain performance data and compare results with competing technologies.

5. In the *electrosorption* process, an electric field controls the adsorption characteristics of activated carbon; this field could also control the adsorption–desorption cycles for a carbon bed used to tract industrial wastewater. Researchers must run performance data at flow rates ten to fifteen gallons per minute to see if this technology is indeed reliable and cost-effective.

While these commercially available electrotechnologies offer attractive solutions to the growing problem of treating industrial wastewater, these technologies, with the exception of biological treatment, are not widely used by industry today. Unfortunately, industry lacks reliable information sources about these systems and their performances; in addition, industry, in general, exhibits great reluctance to use new, costly, and potentially risky technologies. Manufacturers of this equipment should consider providing support for commercial scale demonstration projects of these electrotechnologies and their benefits. These case studies would prove the viability of these electrotechnologies and would encourage industry to accept these products as cost-effective, environmentally sound alternatives to wastewater treatment.

References:

3-1 M. Shepard, "The Politics of Climate," *EPRI Journal*, June 1988.

3-2 A. Lannus, "The Great Heat Pump Debate: Conservation Technology?", *The Electric Heat Pump Future*, Electric Power Research

Institute, November 1989.

3-3 "Environmental, Energy, and Economic Effects of Residential Heating and Cooling", prepared by Entechnology Inc., Chattanooga, Tenn. for Electric Power Research Institute, Palo Alto, Calif. RP No. 2597-19.5, Draft Report, July 1990.

3-4 "Carbon dioxide reduction through electrification of the industrial and transportation sectors," Edison Electric Institute, July 1989.

CHAPTER 4
Electricity's Relationship to U.S. National Security

Iraq's temporary annexation of oil-rich Kuwait in late 1990 illustrates the complexities, complications, and contentiousness associated with the world's growing dependence on energy. Its use by the world's industrialized countries has increased by over 60 percent during the past two decades, and industrial energy use has doubled (see Table 4-1). Developed nations as well as emerging countries rely on energy supplies such as coal, oil, and natural gas as well as hydropower and nuclear energy to power their industries and provide energy services to their citizens.

Table 4-1
Trends in U.S. Industrial Energy Use*

	1890	1910	1930	1950	1970	1990
Total World Energy Use (Terawatts)	1.00	1.60	2.28	3.26	8.36	13.73
Cumulative Industrial Energy Use Since 1850 (Terawatt-years)	10	26	54	97	196	393

* Adapted from: John P. Holdren, "Energy in Transition", *Scientific American*, Sept. 1990, page 160.

The environmental consequences of this ever growing need for energy have become increasingly apparent (see Chapter 3). However, this demand for energy has also generated serious social, political, and economic repercussions. First, these consequences do not honor national border restrictions. Second, in the eyes of some, the environment is a viable and legitimate strategic tool as well as weapon to be used when waging war. And, as recent events in the Persian Gulf illustrated, energy service facilities represented both prime targets and were used as weapons during acts of terrorism and war.

The Global Marketplace Deemphasizes National Borders

The once distinct borders that surrounded each nation have become less pronounced. Past practices of conducting social, political, environmental, and economic activities based on one nation's needs and goals are no longer feasible because these activities often now affect several nations concurrently. In the energy marketplace, this interconnection between nations is particularly strong. As Peter Gleick has noted, "a nation or region bent on protecting its 'security' in the future will have to concern itself as much with the flows of the planet's geophysical capital as it does today with the flows of economic capital."(4-1)

Several examples illustrate the global impacts of energy-related issues. While world energy supplies remain relatively abundant today, prices fluctuate from nation to nation. (Prices in some areas, including the U.S., remain low.) Because the number of oil producing nations continues to shrink, more nations depend on purchasing oil from outside their boundaries. For example, oil currently accounts for well over 40 percent of U.S. energy use and the country purchases a large percentage of this oil from other countries. Future projections indicate the U.S. and many of its allies and international business partners are likely to become even more dependent on imported oil, particularly from producers in the Persian Gulf. This growing dependence on imported oil increases the risk the U.S. (and the world) faces from supply disruptions, price controls, and armed conflicts. These risks damage the U.S.'s economic well-being and its national and energy security.

Reliance on imported oil inflicts a variety of social, political, and economic costs on the U.S. government (beyond basic disruptions in the balance of trade). U.S. allies also feel these impacts in a variety of ways:

1) U.S. foreign policy as well as that of its allies is ever more influenced by the actions of oil rich nations (or threats to those nations). The U.S. faces a weakened negotiating stance—ultimately, the "other side" knows the U.S. needs oil—and this weakness becomes a factor in negotiations. The U.S. did not only send and lose soldiers to liberate Kuwait because of some generic threat to democracy or basic human rights—Kuwaiti oil fields were under attack.

2) When oil supplies become scarce (or perceptions suggest its limited availability), the U.S.'s relationship with its allies can be undermined or at least tested. In the Persian Gulf War, the U.S. not only had to deal with Saddam Hussein and his government in Iraq, it also had to coordinate efforts with each of its allies in the United Nations regarding the war's goals, execution, and costs.

3) An actual supply disruption directly affects the U.S. and world economies as well as limits the ability to meet energy needs. If this disruption coincides with a major defense emergency, then problems escalate.

4) The U.S.'s military preparedness depends on availability of resources such as oil. The U.S. must maintain adequate energy supplies for national security purposes, both emergency and nonemergency.

The environmental impacts of energy sources and supplies are equally complex. A nuclear disaster in one country may first occur there, but as with Chernobyl, its effects may spread rapidly to neighboring nations. Wind currents from the Chernobyl explosion spread radiation far beyond the former Soviet Union's boundaries (e.g., milk shipped from the area to the Scandinavian countries contained abnormal levels of radiation). While the majority of the human suffering and environmental destruction caused by the Chernobyl disaster is contained to one region in the former Soviet Union, the repercussions from this accident reverberate world-wide. As a result of this incident, many nations became more concerned about the potential for additional energy plant disasters in today's Commonwealth of Independent States as well as their own neighbors. In this instance, the energy-related environmental disaster was not deliberate—human error and system design flaws resulted in catastrophic consequences.

The Environment as a Tool and Weapon

There are many examples throughout history of nations manipulating the environment during war time or fighting over its bounty. Nearly 2,400 years ago the Thasians and Athenians fought over control of mineral resources. The U.S. used chemicals to defoliate the jungles in Vietnam so that combatants had a better chance of seeing the enemy. Both Allied and Axis powers regularly bombed dams in World War II and Korea to flood areas and play havoc with water supplies. Finally, water-rich nations (and even individual states in the U.S.) continually threaten to withhold water from more arid countries (or states). In these examples, warring nations (or neighboring states) use the environment as a strategic tool: they fight over resources, attack it, and attempt to co-opt resources from their enemies (or competitors) to gain leverage.(4-1)

However, during the past twenty-five years in particular, nations have begun to use the environment as a weapon during war time; a weapon that has global, not local impacts. Unfortunately, some of the most drastic uses of this "weapon" involve energy sources or supplies. The most recent example is Saddam Hussein's destruction of the Kuwaiti oil fields. In setting hundreds of oil wells on fire, he and his government were not simply "punishing" Kuwait, or even Kuwait and its immediate neighbors and allies (e.g., Saudi Arabia). The smoke and ash pouring into the atmosphere for months affected the global environment and will result in world-wide economic consequences for years to come.

During this most recent war, as well as past conflicts (e.g., the 1967 Six Day War between Israel and Syria), nations targeted power plants and oil fields. In 1981, Israel attacked the Osirak nuclear plant outside of Bagdad to stop Iraqi nuclear weapons research. In January of 1991, the U.S. and its allies first targeted Iraqi nuclear facilities. In these recent examples, energy sources and supplies have become both sound military targets and lethal weapons. Accordingly, they continue to play an important role in a nation's "national security" planning. Analysts can (and do) argue extensively over semantic issues (e.g., should "national security" now be referred to as "natural" or "environmental" security).(4-2)

But facts point to one inescapable conclusion: energy sources and supplies present obvious military targets and represent powerful weapons. These strategic attacks do not simply destroy power plants and disable enemy energy sources, they can also potentially harm nations uninvolved in the current conflict. (Which, at least peripherally, drags them into the conflict and further complicates the political relationships among all parties.) One way to lesson the value of energy as a weapon: nations must become more energy independent and use it more efficiently.

U.S. Energy Dependency: Oil Imports (4-3)

As the Persian Gulf War so ably demonstrated, conflicts in the Middle East can disrupt oil supplies and pose a great national security risk to the U.S. and its allies. So why does the U.S. continue to maintain its dependence on imported oil when it could substitute other energy sources, namely electricity, and reduce this need for oil (and the attendant security risks)? The electric utility industry currently generates very little electricity from oil-burning plants. However, U.S. citizens and its industries still rely heavily on oil-fueled end uses, such as automobiles and industrial processes. If U.S. consumers would switch from oil-fueled end uses to electric-powered end uses, the country could reduce its dependence on imported oil and lessen the very real threats presented by this energy dependence.

Oil Supply Disruption Cost: A Case Study

In 1987, a U.S. Government interagency technical group consisting of staff from the Council of Economic Advisers, Office of Management and Budget, Department of Treasury, and Department of Commerce evaluated the potential effects of temporary oil supply disruptions on the oil market and the domestic economy. The study examined both historical supply disruptions involving net reductions in world oil supplies of about two million barrels per day and hypothetical net reductions in world oil supplies of ten million barrels per day.

In the event of a two million barrels per day net reduction, the Department of Energy (DOE) estimated that drawdown of the U.S. strategic petroleum reserves and some other countries' emergency stocks would sufficiently offset the impact of the disruption over a six-month period. At the time, DOE claimed that moderate supply disruptions would have little effect on world oil prices. This conclusion was verified in late 1990 and early 1991 after Iraq's invasion of Kuwait. In that market reaction, prices tended to rise toward the post-OPEC embargo prices and then dropped again.

The interagency group also analyzed the consequences of major supply disruptions. Table 4-2 summarizes the results. The "major disruption" scenario examined involved a six-month net reduction of about ten million barrels per day in free world oil supplies. The group's hypothesis (and the table) expresses disruptions in terms of world price shocks. Although the supply disruption in this hypothetical example lasts only six months, the price shock persists for a full year after the onset of the disruption, reflecting unwillingness of the oil price to return to its predisruption level.

Table 4-2
The Impact of Supply Disruptions
(1985 $)

	OIL PRICE PER BARREL			BILLIONS	
	Base Price	Possible Price Following Disruption	Difference	Trade Loss	GNP Losses
1990	16→23	39→59	23→36	(-24)→(-71)	(-41)→(-104)
1995	22→28	45→72	23→44	(-35)→(-118)	(-42)→(107)

The "GNP (Gross National Product) losses" and the "trade loss" of the disruption were calculated and reported separately. While both measurements are important, they represent different kinds of economic losses. GNP is a measure of value added, calculated on the basis of the value of domestic production. As such, it does not measure the change in purchasing power or wealth that would be the result of a large increase in the world price of oil and the resulting increased expenditures for oil imports. Nevertheless, the transfer of wealth from oil consumers to oil producers would represent a real economic loss to the U.S., since the U.S. is a net oil importer.

The terms-of-trade effect (trade loss) is relatively easy to compute. It is calculated as the component resulting from: 1) the increase in the price paid for oil imports, plus 2) the deadweight loss (measured by consumers' surplus) resulting from the price shock, and 3) subsequent reduction in oil consumption.

Assessing the Costs of Fuel Switching

As the U.S. has grown increasingly dependent on unreliable sources of costly imported oil, economists and policy analysts have begun cautioning that the market price of imported oil may not reflect its total social costs. For example, switching from oil-burning to coal-fired or nuclear generating plants cannot occur without affecting consumers. Accordingly, many analysts believe utilities must examine the impacts of two nonmarket factors—the environmental hazards associated with power generation and the economic impacts of imported oil—when assessing the price currently paid for oil. These factors affect specific individuals, customer markets, and society as a whole in widely divergent ways.

Thus, to understand the role of nonmarket impacts in demand- or supply-side management decisions, utilities must identify costs and benefits separately for each group. For example, suppose a utility wants to replace an oil-burning plant with a baseload coal plant. If total electricity sales do not change when the utility makes this switch, different air emissions will occur as a result of substituting coal for oil. These air emissions may negatively impact individuals living closely to the coal plant and positively impact towns located further away that were once downwind from the old oil plant. The displacement of oil by coal will eventually lead to a reduction in oil imports, which will benefit the nation as a whole. So, people living around the coal plant enjoy the benefits of reduced oil imports, yet bear the costs of increased air pollution. People living in the rest of the country enjoy the benefits of the reduced imports and bear no costs. Thus, a particular utility strategy may offer acceptable costs and benefits to society as a whole but its targeted customers (those impacted most directly) may find the strategy extremely unattractive. (4-4)

A number of recent studies have examined the issues associated with imported oil and have attempted to quantify the different components of imported oil's externality cost. These components include the nonmarket factors introduced above. Estimates of imported oil's externality cost, also called the imported oil premium, range from about $1 to $60 or $70 per barrel, with most estimates in the low end of the range, from about $1 to $10 per barrel. These nonmarket impacts are not a unique problem associated with DSM or energy production by electric utilities. Externalities occur in every facet of modern economic life. Economists and policy analysts have struggled to measure the issues they represent for years.

Changes in oil imports will result in three different economic impacts. First, a reduction in oil imports applies downward pressure to hold oil prices constant or to reduce oil prices in real terms. Second, import reductions decrease the U.S.'s vulnerability to undependable oil supply sources. The less oil imported, the less costly the impact, if supply reduction (or elimination) occurs. Third, import reductions can result in macroeconomic impacts that could strengthen the U.S. dollar, decrease inflation, and improve the U.S. balance-of-trade position. These macroeconomic impacts are more important in the short run than in the long run.

One useful analysis which supports this hypothesis comes from how the economies of different countries reacted to the 1973 oil embargo. This embargo destroyed all traditional relationships between electricity and economic activity. Based on a study of 183 countries or dependencies, 123 countries achieved a net increase in constant dollars in GNP per capita between 1973 and 1978. [4-5]

The average gain in GNP per capita ($937.40) was obtained with an added expenditure of 1,004 kWh at $37.65, plus 2,091,431 Btu at $10.50 for a total of $48.25. This total is equivalent to an incremental energy cost of $4.05 per $100 of added GNP. This low cost is made possible by the unique effectiveness of electricity in "generating GNP." [4-5] Countries with greater share of electricity use less total energy per GNP dollar. Table 4-3 illustrates the forty-one countries and dependencies with the highest growth in absolute terms.

While there is a wide variation, the pattern shown at the bottom of the table is clear. The growth of electricity use must outpace economic growth by at least two percentage points or progress at a 50 percent faster rate. Whenever limitations are placed on electricity growth, economic growth will be severely hampered.

Table 4-3
Per Capita Increases in Absolute Terms and in Percent, From 1972 to 1978, for the 41 Countries or Dependencies Which had the Highest GNP/Capita Gain. (Source: 4-5)

Countries or Dependencies	1977 U.S. & GNP Per Capita	Average Annual Increase	kWh Per Capita	Average Annual Increase	kec Per Capita	Average Annual Increase
Saudi Arabia	3,630.4	12.1%	193	13.8%	380	6.0%
Bahrein	2,959.1*	27.7%*	1,469	9.6%	2,978	5.9%
Brunei	2,645.8	5.4%	1,035	10.0%	263	1.2%
Norway	1,749.0	3.7%	2,369	2.4%	1,292	2.3%
Greenland	1,639.7	5.2%	570	4.1%	-1,050	-2.8%
Japan	1,300.2	3.5%	904	3.4%	273	1.2%
Belgium	1,276.0	2.8%	990	3.9%	269	0.7%
Canada	1,266.0	2.7%	2,519	3.6%	1,056	1.4%
Germany Fed. Rep.	1,204.7	2.5%	1,204	3.9%	673	2.0%
France	1,148.0	2.7%	1,029	4.7%	329	1.2%
Bermuda	1,117.1	2.3%	348	1.1%	899	5.4%
Austria	1,087.9	3.0%	1,105	4.5%	612	2.2%
United States	1,063.2	2.1%	1,727	3.0%	-103	-0.14%
Martinique	1,026.3	5.6%	217	7.2%	221	4.2%
Hong Kong	994.1	7.4%	455	4.5%	215	2.3%
Denmark	932.0	1.8%	824	3.4%	59	0.2%
Singapore	897.5	6.1%	1,010	9.2%	995	9.1%
French Polynesia	879.3	3.3%	585	8.4%	-268	-4.7%
Netherlands	852.8	2.0%	870	3.6%	-71	-0.2%
Gabon	830.9	4.8%	592	21.7%	860	11.1%
Germany Dem. Rep.	825.0	4.7%	1,424	4.8%	1,090	2.8%
Iceland	784.6	1.9%	3,625	6.2%	1,553	3.7%
Finland	719.4	2.1%	1,397	3.7%	550	1.7%
Romania	659.8	9.0%	909	6.7%	807	3.6%
Iraq	649.6	8.2%	276	12.2%	-117	-2.4%
Kuwait	619.1	0.75%	1,425	4.75%	2,640	-4.25%
Czechoslovakia	609.0	4.05%	1,021	4.05%	731	1.75%
Poland	602.0	5.5%	977	6.1%	1,066	3.5%
Yugoslavia	565.7	5.0%	691	6.25%	445	3.55%
Sweden	557.8	1.0%	1,522	2.65%	586	1.35%
Venezuela	537.7	3.7%	490	5.3%	349	2.1%
Bulgaria	534.0	5.6%	1,048	5.65%	1,170	4.05%
Australia	522.8	1.2%	1,426	4.6%	1,009	2.65%
Hungary	514.0	4.7%	839	6.05%	450	2.3%
Greece	510.4	7.0%	559	5.5%	371	3.5%
Guadeloupe	506.2	3.5%	268	9.65%	179	4.7%
USSR	479.0	4.05%	1,041	4.45%	1,074	3.5%
United Kingdom	463.4	1.75%	453	1.75%	-179	-0.55%
Ireland	441.9	2.45%	799	5.05%	28	0.15%
Republic of Korea	433.7	8.95%	468	14.25%	585	9.5%
Spain	415.8	2.25%	754	5.65%	657	4.6%
Average	937.3	4.06%	1,004.0	6.13%	454.8	2.23%

* Not included in averages

An analysis of the data in this study reveals that electricity provides most of the total added energy use. According to Felix,[4-5] 83.5 percent of the added energy is provided by electricity and 16.5 percent by non-electric sources. Seven of the countries listed in Table 4-3 achieved substantial net economic growth while reducing their total energy use per capita. This growth reflects greater electrification, which displaces inefficient fossil fuel use.

Clearly, electricity's ability to reduce foreign oil dependence rests, in part, on the diversity of existing and potential generation sources as well as the added assurance of continuous availability provided by power pools. The greater the diversity of energy supplies, particularly electricity, the more efficient a nation's energy use becomes. In addition, as Felix[4-5] observes, "...products which require a greater share of electricity are more sophisticated, more innovative, more valuable. They are lighter. They save materials and energy. They are in greater demand. They are more productive. They create more jobs."

In advocating U.S. energy independence, some policymakers have argued that the external costs of nuclear plants are less than those of fossil-fuel-powered plants. Analysts estimate the environmental costs for the nuclear fuel cycle based on several factors, including routine emissions from reactors and other parts of the fuel cycle, occupational radiation exposure, nonradiation related accidents, and the expected costs of nuclear accidents and sabotage. In an illustrative analysis for the National Science Foundation (NSF)[4-6] analysts estimated the externality cost at 0.10 mill per kilowatt-hour of nuclear electricity generation. The most significant portion of this cost can be attributed to occupational hazards and the risk of sabotage. The social costs of routine low level radiation was considered negligible.

Increasing this value to $1,000 per full-body rem for exposure within fifty miles of the plant, based on an interim standard proposed by the Nuclear Regulatory Commission, would increase the social cost to 0.15 mill per kilowatt-hour. The calculations in the NSF report of nuclear accident probabilities and consequences are based in part on those determined in the Rasmussen report.[4-7] A subsequent study by Ford-Mitro[4-8] estimated that the average loss due to nuclear accidents may be underestimated by a factor of 500. In this most pessimistic case, considered highly unlikely, the social cost of nuclear power generation would be approximately 1.6 mills per kilowatt-hour. Thus, a range of 0.10 to 1.6 mills per kilowatt-hour for nonmarket impacts of nuclear power seems reasonable.

Environmental costs for fossil fuel plants result primarily from health effects, material damage, acid deposition, and visibility reductions due to emissions of sulfur oxides, nitrogen oxides, and particulates into the atmosphere. A study for the National Academy of Sciences[4-9] estimated the cost of SO_X and particulate emissions for two representative coal plants to assess adverse impacts on human health and the environment. Then, analysts translated these impacts into dollar terms. [4-3]

For a remotely located hypothetical plant in western Pennsylvania or West Virginia, analysts estimated the cost of these nonmarket impacts to be 21 cents per pound of SO_X (1975 dollars), with credible values ranging from 9 cents to 37 cents per pound. For a plant located 35 miles outside of New York City, they estimate the social costs to be 55 cents per pound, with a range of 21 cents to one dollar per pound. [4-3]

The NSF study includes estimates of the social cost of NO_X and particulate emissions, based on extremely crude calculations of the damage these pollutants cause to human health and materials. For NO_X, they estimate the social costs to range from one to ten cents per pound of emissions, with a nominal estimate of three cents per pound. The cost of particulate control equipment is included in the cost of new power plants, and the cost of residual emissions is small in their estimates. Thus, they do not assign a separate social cost to particulate emissions.[4-3]

Both these reports emphasize the illustrative nature of the estimates they contain. However, they are useful for determining a broad range within which the nonmarket impacts of environmental hazards resulting from electricity generation may lie.

The Gulf War has undoubtably had tremendous impacts on this scenario although, to date, analysts have not updated a study like this based on the post-war situation.

Conclusions

Energy clearly plays an important role in U.S. national security—both as a strategic tool and a potential weapon. Increasing electrification enhances energy efficiency as well as mitigates the potential for the U.S. (as well as the remaining countries in the world) to be "held hostage" to a dependence on energy imports. In addition, industries that substitute new electrotechnologies for older technologies reduce the potential for negative environmental consequences.

References:

4-1 Peter Gleick, "Environment and Security: The Clear Connections", *The Bulletin of the Atomic Scientists*, April 1991.

4-2 Daniel Deudney, "Environment and Security: Muddled Thinking, *Bulletin of the Atomic Scientists*, April 1991.

4-3 *Cost-Benefit Analysis of Demand-Side Planning Alternatives*, EM-5068, Electric Power Research Institute, November 1983.

4-4 C. W. Gellings, et.al., *TAG™ Technical Assessment Guide Volume 4: Fundamentals and Methods, End Use*, Electric Power Research Institute, P-4463-SR, Volume 4, August 1987.

4-5 F. Felix, "Our Top Priority: Expanded Electrification Will Substantially Reduce Oil Use, While Propelling Economic Recovery", *The Economy and Electricity Conference 1980*, Gibbs & Hill, Inc., New York, New York, 1980.

4-6 S. M. Barrager, B. R. Judd, and D. W. North, *The Economic and Social Costs of Coal and Nuclear Electric Generation*, Report prepared for National Science Foundation (Washington D.C.: U.S. Government Printing Office, 1976).

4-7 *Reactor Safety Study*, U.S. Nuclear Regulatory Commission, Main Report, October 1975.

4-8 *Nuclear Power Issues and Choices*, Nuclear Energy Policy Study Group (Cambridge, Mass.: Balinger, 1977).

4-9 D. W. North and M. W. Merkhofer. "Analysis of Alternative Emissions Control Strategies", *Air Quality and Stationary Source Emissions Control*, A report by the Commission of National Resources, National Academy of Sciences (Washington, D.C.: U.S. Government Printing Office, 1975).

CHAPTER 5

Electricity's Impact on Family Life

The invention of electricity irreversibly altered family social structure in general and women's lives in particular. Prior to the Industrial Revolution, both men and women performed "domestic work." That is, because a majority of the families lived an agrarian lifestyle, daily tasks centered around the home and farm, which formed the center of production. Every member of a family held a place in the household and performed specific chores. Men and women held equal levels of responsibility and both parties were valued equally for the work they did—their contributions to the household. Most tasks—whether household or farm chores—demanded a great deal of physical energy and large blocks of time from both men and women.

Everyone—men and women alike—"worked," and worked very hard! Each family member contributed to the welfare of the family. Men and women, from an early age, worked for money, for credit in barter, or for room and board. Each type of labor performed by both sexes was considered work, even if hard currency did not change hands. Many tasks were "pooled" by a community of households. For example, a community often maintained a gristmill that all could use to process grain or it would band together to accomplish tasks such as constructing houses and barns, making quilts, or harvesting crops. Individual households as well as the community understood the absolute necessity of working together to survive and prosper.

The first utilities—piped water and gas—appeared in the 1830s, however, only in the homes of the urban dwelling upper middle class and rich at this early stage. Prior to this date (and for many households long after), water for chores such as cleaning and cooking had to be extracted from wells or other water sources by hand pumps and then carried to its point of use. If heated water was required, someone in the family chopped the wood or shoveled the coal that fed the stove or fireplace that heated the water. Then, the women and girls washed and ironed the clothes by hand (the iron was heated by placing it on the same stove). Because stoves and fireplaces produced so much soot and grime, the end of the wash day was spent cleaning up the ashes spread by the fire. Obviously, many of these basic household chores were extremely labor intensive and the introduction of these first utilities eased workloads considerably. By the end of the nineteenth century, these basic utilities spread to all but the very poorest homes.

Electricity's Introduction into Domestic Life

Electricity was first commercially generated and distributed during the period 1879–1881. In the beginning, cities and towns used it only for artificial illumination, especially streetlamps. As electric technologies evolved, inventors and utilities began understanding its potential applications in the domestic "workplace" as well as in newly emerging industries. However, for many years, technological limitations restricted the use of multiple electric appliances. For example, when used to power other appliances such as early electric water heaters, utilities had to provide a major commitment of secondary distribution reinforcement so as not to dim the lights for miles around. (5-1)

By 1890 Thomas Edison had created small motors that could power household appliances. This advancement paved the way for numerous applications. The electric iron was actually the second major residential appliance (after lighting) to achieve widespread use. In 1908 the first Hoover vacuum cleaner was introduced. Electric washing machines and refrigerators were available soon afterwards. Refrigerators were not effectively mass produced until the 1930s.

By the 1920's, two-thirds of U.S. households used electricity to power lights and appliances. As with most new inventions, households needed capital to purchase these conveniences. Early in their introduction, only wealthy and upper middle class households could afford new electric appliances. As sales increased, costs of production and distribution declined, and new appliances became affordable to middle class and poorer households.

Rapidly evolving household technologies included: (5-1)

- Utilities (central water supply, electricity, natural gas, sewage)

- Appliances (such as electric irons, mixers, vacuum cleaners, stoves, ovens, washing machines, and light fixtures)

- Convenience foods (such as perishable items that could be stored in refrigerators)

In addition to household use, many of these technologies contributed to a burgeoning service industry (e.g. laundries).

As shown in Table 5-1, even in 1940, many technologies were not used widely in U.S. homes. By 1987, however, more than one-half of the 90.5 million households in the U.S. owned all of the major appliances. Table 5-2 details the major electric appliance use as a percent of households. As the next section discusses, these appliances displaced or substantially altered many traditional household functions and activities.

Table 5-1

Trends in Labor-Saving Technology (5-2)

	Percent	
Facility	1940	1960
Bathtub or shower	61	88
Wood, coal or oil stoves	45	5
Wood or coal for heating	78	16
Electricity	79	99

As electric technologies evolved, electricity users began to recognize electricity's ability to provide a clean energy supply and contribute to cleaner living conditions. Electricity use eliminated the dirt and grime associated with burning coal and wood as well as facilitated food preparation, storage, and clean up. Early medical researchers increasingly recognized the link between disease and unsanitary conditions as well as the ways electricity use could solve these problems. From a cultural perspective, Americans linked cleanliness with moral decency and good behavior. Thus, as washing garments became easier, people changed them more frequently and machines scrubbed them in hot water using strong detergents and starch. (5-3) Today, Americans are known worldwide for their attention to personal cleanliness, including routine bathing and frequent clothes changes.

Table 5-2

U.S. Household Electric Appliance Use: 1987

Type of Appliances Used	Percent of Households
Television Set (color)	92.7
Television Set (b/w)	35.8
Clothes Washer (automatic)	73.3
Range	56.8
Oven	56.6
Clothes Dryer	50.7
Dishwasher	43.1
Window or Ceiling Fan	46.2
Microwave Oven	60.8
Water Heater (for one household's use only)	33.7
Air Conditioner (room)	30.8
Electric Blanket	30.0
Air Conditioner (central—for one household's use only)	32.5
Freezer (not frost-free)	23.0
Humidifier	14.6
Freezer (frost-free)	11.7
Waterbed Heater	13.9
Dehumidifier	10.0
Whole-House Cooling Fan	9.5
Swimming-Pool/Jacuzzi/Hot-Tub Heater	.7

Refrigerators	Percent of Households
1	86.2
2 or more	13.6

In addition to its direct benefits, electricity produced many secondary effects, including an expansion in manufacturing. For example, electricity powered many central heating systems: it provided a source of fuel as well as powered the controls, fans, and pumps. In 1929, the average home used 484 kWh per year of electricity. By 1971, the average use grew to 7,385 and by 1985 increased to over 9,000 kWh. In 1971 alone, manufacturers sold 575,000 electric space heating units, bringing the U.S. total to 5.2 million. [5-4]

Social Consequences of the Introduction of Electricity

The early 1920's represented a period of rapid social and political change in the U.S. Post-World War I optimism, combined with rapid technology advancements, provided major catalysts for this change. The growth of electricity use spurred many of these social changes. The consequences of these changes are felt even today, both in domestic and industrial arenas.

Domestic Impacts

The introduction of electric-powered equipment in both homes and businesses irreversibly altered the domestic roles of men and women. The traditional agrarian lifestyle required that both men and women perform domestic duties. However, as electric-powered technologies evolved, their use spread to towns and cities throughout the U.S. Business and industry expanded nationwide and produced jobs that lured many people to towns and cities from family farms. Chapter 8 discusses this spread in greater detail.

As a result, families often sent the head of a household (usually men) and younger children (girls and boys) off to work each day, leaving other women family members to shoulder the domestic duties. Thus, in the home, the split began to widen between "men's" and "women's" work. Thomas Edison and other pioneers of the industry encouraged this split (albeit unaware of the larger social consequences) by developing electric appliances for the home.

Utilities then targeted many of these appliances at women because they were perceived to relieve women (primarily) of the drudgery of certain household chores. Markets grew rapidly, particularly as marketing campaigns for them increased. Beginning in the early 1920s, appliance manufacturers and electric utilities implemented concerted marketing efforts to bolster electricity (and new appliance) use. They employed women as home economists to demonstrate the benefits of these new "convenience" tools. Women users were regarded as the advance guard of the appliance revolution. (5-5)

Using these new appliances reduced the amount of time women spent on certain household tasks; the work they did shifted from tasks of production to those of maintenance. For example, canning, processing, and refrigeration techniques provided prepackaged food items. The housewife no longer had to rely solely on the foodstuffs produced in her garden. Table 5-3 depicts the time spent on housework during several different decades.

The advent of electric powered appliances changed housewifery: technology reduced the work process to simpler levels. It did not eliminate the household work; in fact, it often *added* to a woman's housework. Before these advances, the housewife in all but the poorest families often sent work out (to laundresses, for example) or employed help in the home—young unmarried women relatives, hired help, indentured servants, or, prior to the late 1860's, slaves. These new electric powered "labor saving" devices often replaced these extra helpers, and left the woman of the house to deal with the chores alone.

Table 5-3

A Comparison of Data on Housework Hours [5-6]

Study, and Country Carried Out In:	Date	Average Weekly Hours of Housework
Rural Studies		
Wilson	1929	64
U.S. Bureau of Home Economics	1929	62
Cowles and Dietz	1956	61
Urban Studies		
U.S. Bureau of Home Economics	1929	51
Bryn Mawr		
(i) small city	1945	78
(ii) large city	1945	81

Industrial Impacts

The industrial impacts of electric power fell into two familiar categories: the "demand" and "supply" sides. On the demand side, consumers wanted more electric appliances (e.g., washing machines) and the goods and products produced by electric powered equipment (e.g., machine woven cloth, processed foods). On the supply side, newly created industries needed workers to produce these appliances, goods, and products.

Expanding industrialization placed growing financial pressures on families, particularly those living in urban areas. A farm could continue to support a family as it had done for hundreds of years. However, as urban areas grew, families living in these areas needed more and more hard currency to pay for the necessities of life (e.g., food and clothing). Basic survival as well as the drive to purchase the ever-evolving selection of new appliances (and pay for the electricity to power them) prompted many family members, particularly young women, to look outside the home for work.

Industries needed to fulfill the demand for their products and often the biggest problem was maintaining an adequate work force. More and more women who lived in and near urban areas wanted (and needed) to work in different industries. However, social pressures frowned on women who left their homes. Industries were constantly searching for new workers, and young women often fit the requirements. In particular, industries such as the textile industry discovered that New England's farmers often had one or more daughters who were willing to leave the farm and work for wages. However, as the years passed, salaries and working conditions deteriorated.

Because social pressures still stressed that women belonged at home, employers used this excuse to pay them substantially less for their efforts than men—it was assumed that their family would take care of them. In addition, industry did not generally train women for more skilled positions (for fear the training would take them even further from the home and its responsibilities). Thus, they remained at unskilled, low paying positions.

Post-World War II demographic changes strongly influenced the adoption of household electrical appliances. During this time period, important demographic changes, particularly the rising participation of women in the labor force and steady reductions in household size, occurred. These factors contributed to rising household incomes as well as the demand for labor-saving devices.

The Dilemma Between Work And Home

Society discouraged women's move into the workforce, however many millions of families needed the additional earnings of these workers to survive. Despite these needs, U.S. society still promoted the ideology that "a woman's place is in the home." By the 1920s, many social commentators were decrying the reduction in household work, fearing that "too many women were dangerously idle." In 1911, an editor of the *Ladies' Home Journal* opined that "What a certain type of woman needs today more than anything else is some task that would tie her down. Our whole social fabric would be better for it." (5-7) This collision of ideals between social perceptions of womens' roles and individual actions still challenges U.S. society today.

Another consequence of these conflicting ideals: through the decades (and even today), women who work outside the home still perform a majority of the household tasks. In a study conducted in the United Kingdom and published in 1983, J. I. Gershunny observes that it was not until the 1960s that minutes of domestic housework per average day actually fell among women who were not employed full-time outside of the home (see Table 5-4). Table 5-5 illustrates the hours per week spent on household tasks, by employment status. Current studies in the U.S. show that, despite the growing number of dual income families (both husband and wife working full time), women still perform over 80 percent of the household chores. These tables also illustrate the point made earlier in this chapter): while electric appliances often simplify domestic chores, they do not eliminate the time it takes to perform them.

Table 5-4

Housewives' Domestic Work Time, Per Day (5-8)

YEAR	MINUTES
1937	410
1958	425
1963	450
1970	400
1975	350

Table 5-5

Hours Per Week of Housework Time in the Tasks of Household Work by Residence/Social Class for Nonemployed Women, 1924-31 and by Employment Status for Urban Women 1965-66 (5-9)

	RURAL 1924-28 Nonemployed (N=808)	College Educated 1930-31 Nonemployed (N=692)	URBAN 1965 - 66 Nonemployed (N=357)	Nonemployed (N=334)
All household tasks	51.6	48.1	55.4	28.2
Food activities:	22.8	15.1	16.0	8.2
Preparing meals	15.4	10.6	10.9	5.4
Meal cleanup	7.4	4.5	5.1	2.8
Clothing and house				
Hold linen care	11.5	7.9	8.7	3.9
Laundry	5.5	3.4	6.4	2.9
Mending and sewing	6.0	4.5	2.3	1.0
Home care	9.6	7.4	10.2	5.0
Cleaning	7.5	6.0	8.4	4.6
Other	2.1	1.4	1.8	.4
Family care	4.1	9.8	9.7	3.3
Shopping and				
home mgmt.	2.4	5.2	5.2	3.9
Other	1.2	2.7	5.6	3.9

One final thought about electricity's impact on family life, particularly women's roles: it is interesting to note that women have always worked precisely in those industries most effected by new electrotechnologies. For example, according to a survey published in 1982, women comprise 96 percent of typists, 94 percent of bank tellers, 93 percent of telephone operators, 81 percent of clerical workers and 71 percent of retail workers. [5-10] Each of these jobs require workers to use electrotechnologies to perform their daily tasks (computers, electric typewriters and cash registers, and copy machines, to name a few). In another industry—cigar making—women displaced skilled male workers (who made cigars by hand) when new electric machines were introduced and took over manual making activities.

References:

5-1 P. Bereano, et. al., "Kitchen Technology and the Liberation of Women from Housework," S*mothered by Invention: Technology in Women's Lives*, W. Faulkner and E. Arnold (editors), Pluto Press Ltd., London, 1985.

5-2 J. P. Robinson and P. E. Converse, "Social Change Reflected in Time Use," in A. Campbell and P. E. Converse (editors), *The Human Meaning of Social Change*, New York: Russell Sage Foundation 1972, p. 48.

5-3 R. S. Cowan, et. al., "Clean Homes and Large Utility Bills: 1900-1940, Families and the Energy Transition" (J. Byrne, et al editors), *Marriage and Family Review*, Volume 9, Nos. 1/2, Fall 1985.

5-4 Statistical Report, Sales, *Electrical World*, March 15, 1972.

5-5 C. Frederick, *Selling Mrs. Consumer*, 1929.

5-6 A. Oakley, *The Sociology of Housework*, Martin Robertson, 1974.

5-7 *Ladies' Home Journal*, October 1911.

5-8 C. J. Haddad, "Technology, Industrialization and the Economic Status of Women," *Women, Work, and Technology*, B. Drygulshi Wright, et. al., editors, The University of Michigan Press, Ann Arbor, 1987.

5-9   J. Vanek, "Household Technology and Social Status: Rising Living Standards and Status and Residence Differences in Housework," *Technology and Culture*, The University of Chicago Press, Volume 9, Number 3, July 1978.

5-10   Labor Force Statistics I, U.S. Department of Labor, September 1982.

5-11   *Women Have Always Worked: A Historical Overview*, Alice Kessler-Harris, The Feminist Press, Old Westbury, New York, 1981.

5-12   *A Woman's Work is Never Done*, Caroline Davidson, Chatto & Windus, London, 1982.

CHAPTER 6
The Links Between Electricity, Energy Efficiency, and Economics

Electricity, energy efficiency, and economics are inextricably linked. During the economic expansion of the 1980s, electricity growth paralleled GNP growth while real electricity prices declined 16 percent and overall energy efficiency improved 11 percent. Analysts project that the continued productivity and energy efficiency improvements offered by electrification will cause the rate of increase in primary energy consumption to shrink, by 2010, to less than half of the annual rate of U.S. GNP growth. For these and other reasons, electricity use—which today accounts for 36 percent of U.S. energy consumption—is expected to grow to 42 percent by the year 2000.(6-1)

At the same time, the wise use of electricity through both naturally occurring and utility-induced end-use efficiencies will reduce the demand for electric energy by nearly 15 percent in the year 2000. Further potential exists to reduce the demand for electric energy an additional 18 percent, resulting in a total of 38 percent. Thus, electricity plays a crucial role in improving this nation's energy efficiency and, as this chapter will show, it facilitates the efficiency of the U.S. economy and the economies of individual businesses.

The energy crisis of the 1970s convinced the world to use less energy—and it did. In fact, the U.S. cut back to the point where demand growth essentially leveled out for the first time in a century. Today, the world still needs to consume energy prudently; however, since electricity increasingly drives productivity and innovation, simply *using less* will not help industry compete into the twenty-first century. The conservation ethic of the coming years is *energy efficiency*—making the energy we do use go further.

Increasing efficiency *at the end use* is the key to the long-term prosperity of utility customers, the U.S., and utilities themselves. The customer benefits are obvious: at the very least, end-use efficiency reduces energy bills. Energy efficient technologies also raise productivity and reduce waste. In addition, many new efficient electric technologies can improve product quality, facilitate the production of new products, and improve a business's working environment. The net effect of these improvements: energy efficiency reduces capital requirements and lowers production costs, thereby increasing the competitiveness of individual businesses and entire industries.

These advantages spread throughout the country, supplying the nation with economic benefits as well as increasing overall U.S. competitiveness in world markets. In addition, increases in energy efficiency offer the most cost-effective near term approach for reducing long-term environmental concerns. Electricity, with its unique versatility, flexibility, and efficiency, is the best overall tool for improving local, national, and global competitiveness and quality of life.

But what about the prosperity of utilities themselves? Selling energy is presently the central concern of the utility business, and one might think that energy efficiency could threaten sales. In the short-term, this concern may be valid. However, more and more utilities are taking a longer view: a utility's fortunes rise and fall in concert with those of its customers. Energy efficiency can turn an industrial plant's marginal losses into profits, keeping a valuable customer in business. Helping customers with such problems not only secures—and often increases—sales in the long run, but also improves customer relations.

Energy Efficiency and Its Link With Electricity

Energy efficiency is a measurement of *output achieved* per unit of energy input (e.g., the amount of light a lamp produces per watt of electricity). The term often refers to the performance of specific end uses or energy services such as lighting, heating, cooling, and motor drives. Because of today's technological advances, particularly in electricity-using equipment, all consumers can optimize energy efficiency by replacing, upgrading, or maintaining energy-using equipment.

Electricity offers superior energy efficiency: it is the most highly organized form of energy available, with virtually 100 percent of its energy convertible into useful work. Other advantages of electricity—specifically in manufacturing applications—include: (6-2)

- *Energy intensity.* Electricity can deliver higher temperatures and greater energy intensity than fossil fuels. It can also generate heat inside an object.
- *Speed of production.* Electricity's high energy intensity offers industries greater production output in a shorter period of time. This benefit reduces unit costs by spreading the costs of labor, overhead, and interest on capital over a larger production volume.
- *Precise control.* Electricity can be more precisely controlled than conventional thermal processes; it can be applied at defined points, for specific periods of time, and in exact amounts.
- *Cleanliness.* Electricity is clean at the point of use; it involves no fumes or residues which can damage materials being processed. Its use relieves industry of the burdens

associated with on-site fuel combustion (e.g., costs of fuel-handling equipment and environmental control equipment).
- *Flexible generation base.* Electricity can be generated from a variety of fuels as well as by renewable resources. As a result, it provides a more flexible generation base than conventional combustion equipment.

For these and other reasons, electricity use—which today accounts for 36 percent of U.S. energy consumption—is expected to grow to 42 percent by the year 2000.

Components Of Energy Efficiency

In some cases, discussions of "energy efficiency" actually refer to other related concepts, such as energy conservation, productivity, and economics. For example, energy efficiency does not specifically define how intensively equipment is used, its costs, or the economic benefits it offers. However, the adoption of more energy efficient technologies directly impacts these and other significant energy policy concerns.

For example, Figure 6-1 presents a hierarchy of national energy-related goals and shows the broad benefits that energy efficiency can provide. Beginning with a basic goal of increasing the national standard of living, policymakers can derive three related national policy goals—improve economic efficiency, maintain national security, and enhance environmental quality.

A 1987 public opinion poll demonstrated the importance of these national energy-related goals to the general public.[6-3] In this poll, 62 percent of the survey respondents believed that the U.S. is "likely" once again to experience energy shortages similar to those in the 1970s that threatened our national security. Fifty percent indicated that the nation's energy strategy should include the goal of reducing energy use. While 39 percent believed that the U.S. should concentrate on finding new ways to produce more energy, 53 percent opposed the relaxation of environmental standards. In a more recent survey (1991), 64

Figure 6-1. Energy efficiency addresses several key national needs.

percent of Americans said they would *sacrifice* economic growth for the sake of the environment. (6-4)

Upgrading and Electrification: Two Paths To Energy Efficiency

The U.S. can achieve energy efficiency gains in two ways: by upgrading or electrifying new processes. To upgrade existing systems, customers can replace currently installed electric equipment with more efficient models. Electrification can occur either by expanding electricity use in new applications or replacing fossil-fueled equipment with electric equipment in existing applications.

Both of these methods of improving energy efficiency will produce environmental benefits, and cost and/or energy savings. For example:

- Replacing fossil fuel burning equipment with more efficient electric technologies would reduce CO_2, SO_2, and NO_X emissions; all are associated with acid rain. (6-5), (6-6)

- Substituting electric vans for gasoline-powered vans operating under similar driving conditions would release 20 to 55 percent less CO_2 into the atmosphere. In addition, emissions of VOCs and CO would drop by 97 percent or more. (Emission estimates for gasoline-powered vehicles include both tailpipe emissions and emissions associated with refinery operations and fuel distribution. In that EVs do not emit pollutants, all emissions associated with EVs are attributable to the power plants that provide electricity for their operation.) [6-7]
- Conventional gas furnaces emit approximately the same level of CO_2 as do electric heat pumps. However, new high efficiency electric heat pumps produce 20 percent less CO_2 than do advanced gas furnaces. Beyond CO_2 reductions, improvements in electric vapor-compression equipment could hasten an end to the use of CFCs in refrigeration and cooling applications, thereby reducing CFC emissions and ozone depletion concerns. And improved electric heating, cooling, ventilating, cooking, water heating, and other building energy technologies could reduce the creation of indoor air pollutants as well as improve the methods of handling them. Thus, these improved heat pumps address the ever-growing health concerns surrounding indoor air quality.

Energy conservation

Energy conservation refers to the process of *reducing energy consumption*. These steps can include increasing energy efficiency or reducing energy use, either by voluntary or mandated means. To many consumers, the concept of "energy conservation" has taken on a pejorative connotation. They think conservation means "warm beer and cold showers." As the manufacturers of energy efficient appliances know, this image is outdated and incorrect. However, to counter this perception, they often promote a product's "enhanced energy efficiency"—its ability to provide increased services with lower energy demands.

Analysts can measure levels of energy conservation by comparing past with present kilowatt-hour or Btu consumption or estimating future energy savings potential. Consumers can take several steps to conserve energy, including investing in capital improvements (e.g., improving home insulation or purchasing energy efficient appliances) and changing energy use behavior (e.g., setting back thermostats in the winter).

U.S. energy policy regarding energy conservation varied widely during the 1960s, 1970s, and 1980s. Policies differed largely due to short-term factors such as the political climate of the moment. Each U.S. president promoted different (often diametrically opposed, in fact) energy policies and each faced national and international crises involving energy shortages.

Energy productivity

Energy productivity measures the level of *economic value* produced per unit of energy input. Electricity fundamentally enhances energy productivity. Energy productivity rises when existing energy services (e.g., lighting, heating, motor drive) improve in efficiency. In addition, energy productivity, although a narrower concept, is related to economic efficiency. The enhanced energy productivity of new, energy-using technologies (e.g., telecommunications, automation/robotics, information processing) can boost the economic efficiency of product production.

Figure 6-2 shows the energy productivity of the U.S. economy since 1960. Between 1960 and 1972, energy productivity increased at an average annual rate of 0.3 percent; since 1972, the annual rate of improvement has been 2.0 percent per year. These energy productivity trends occur because of a number of factors:

- Growing energy efficiencies of appliances, buildings, and systems
- Initiating short-term changes in energy control and energy management systems

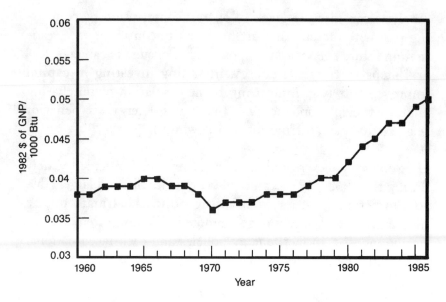

Figure 6-2. Energy productivity in the U.S. economy since 1960.

- Shifting populations, resulting from people migrating from the snowbelt to the sunbelt, lead to reductions in energy use
- Changing the composition of industry from energy-intensive manufacturing to less energy-intensive services
- Increasing energy efficiency of industrial equipment, automobiles, trucks, and airplanes, and electrification of transportation systems

The Economics of Energy Efficiency

Economics play a crucial role in energy efficiency policy, since energy prices and capital costs both affect the level of energy use. In turn, demand for electricity depends on three economics-based factors: 1) the level of economic activity, 2) the price of electricity when compared with alternative energy forms, and 3) the availability of high quality, efficient electricity consuming technologies. Each of these interrelated factors determines the return on investment (ROI) of conservation and of investments in new energy efficient technologies.

The ROI in energy efficiency is highest when energy prices are relatively high and capital costs are relatively low. High energy prices provide the financial incentive to use existing energy equipment in a more efficient manner through greater maintenance or increased use. High energy prices also make new, more energy efficient equipment more economically attractive. The decision to replace existing equipment is made even easier when capital costs are low, because financing costs are lower. Low capital costs (i.e., low interest rates) spur investment in new plant and equipment.

When both energy prices and capital costs are relatively low, the ROI drops. Under these conditions, businesses usually have little financial incentive to focus on energy efficiency measures. However, when capital costs are relatively low, there is an incentive to invest in new equipment. If that new equipment happens to be more energy efficient, energy efficiency improvements will occur with the natural turnover of the capital stock.

ROI is relatively high when both energy prices and capital costs are moderately high—a set of economic conditions similar to those that occurred following the OPEC oil embargo in 1973 and again in mid–1990. Under these conditions, consumers have a substantial incentive to invest in low-cost efficiency measures such as caulking, weatherstripping, and other forms of weatherization.

The financial incentive to invest in energy efficiency is lowest when energy prices are relatively low and capital costs are high.

Economic Efficiency: The Relationship Between Electricity Use and GNP Growth

Researchers generally measure economic efficiency based on the output of several factors of production, including capital, labor, energy, materials.[6-8] Analysts can express the relationship between electricity use and economic growth in aggregate terms: total available electricity and U.S. GNP, expressed in constant dollars. When they use standard statistical techniques to measure the relationship between electricity use and GNP, certain regular features appear. Perhaps the most significant characteristic is the stability, over appreciable segments of time, in the growth rate between electricity use and GNP (Figure 6-3).

Historical trends toward electrification in the residential, commercial, and industrial sectors illustrate electricity's key role in U.S. economic growth. Electricity's share of the residential energy market has grown from about 10 percent in 1960 to almost 30 percent in 1985. During that same period, the natural gas market share has remained roughly constant (about 50 percent) while oil's share has fallen (from about 30 percent to about 10 percent). The commercial sector exhibits similar electrification trends: between 1960 and 1986, electricity's share of the commercial sector energy market grew from about 40 percent to over 60 percent. During this same period, fossil fuel's share (including natural gas and oil) declined by roughly the same proportion that electricity's share grew. The industrial sector has also experienced steady electrification. By 1981, electricity's share of the industrial sector market reached 17 percent, almost doubling in

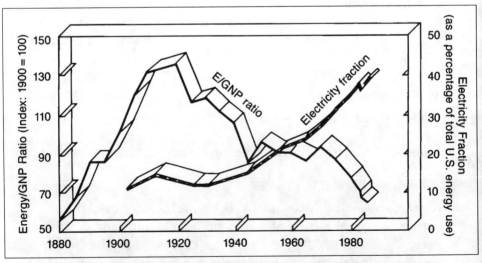

Figure 6-3. As electricity use has increased as a fraction of total U.S. energy use, the amount of energy consumed per unit of economic output–Gross National Product (GNP)–has decreased.

share from the 9 percent level in 1960. Market share forecasts for the industrial sector indicate that electricity's share could be as high as 20 percent by 1995.

Figure 6-4 illustrates the historical correlation between average electricity use and GNP. It uses lines of regression for four time periods in the twentieth century; each period is marked by an approximately stable linear relationship, which, though showing some annual fluctuations, indicates a strong tendency toward a constant incremental intensity of electricity use within each period.

This historical record exhibits only a few changes in the slope and level of the regression line. Clearly discernible changes in the line's slope occurred following World Wars I and II. The Great Depression did result in a shift in the level of the regression line upward, however no change in the slope occurred. During this period, GNP decreased relatively more than did electricity use.

The Post-World War II period (1947 to 1983) marked a time which witnessed the growth of both electricity use and the GNP. Figure 6-5 illustrates this relationship for the Post-World War

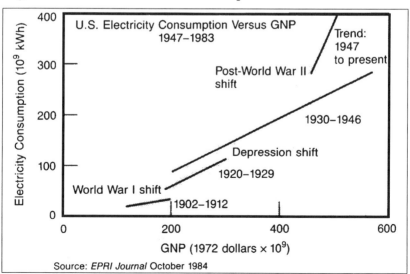

Figure 6-4. U.S. electricity consumption versus the GNP.

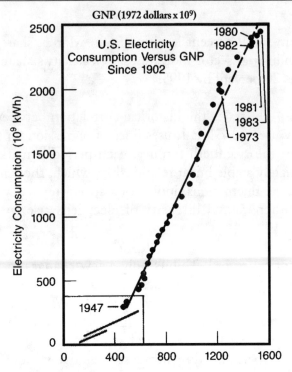

Figure 6-5. U.S. electricity consumption versus GNP, post WWII period.

II period. While the relationship persisted through much of the period, since the mid-1970's analysts have observed a break in the pattern: the ratio between electricity use and the GNP has been decreasing. Some researchers believe that although observations for the most recent years fall below the trend line, this shift merely demonstrates a tendency for individual years to exhibit a cyclical pattern around the long-term trend line. Suddenly, a number of trends once associated with electricity use altered, which led economists to question the future relationship of electricity use and GNP growth.

The positive impact of electrification can also be observed in a global context. A comparison of GNP per capita with electric generation capacity per capita for over 100 nations representing 4.7 billion people shows a direct relationship between electricity use and prosperity: the higher the per capita consumption of electricity, the higher the per capita income, and vice versa.

Current U.S. Economic Efficiency

Economists apply three major statistics to measure economic efficiency within the U.S. economy—output per hour worked (labor productivity), output per unit of capital (capital productivity), and a weighted average of the two (multi-factor productivity). Unfortunately, recent statistics show that economic efficiency is lagging in the U.S. economy, when compared with other industrial nations.

This efficiency drop raises concerns about the long-term competitiveness of U.S. manufacturing and business sectors. Labor productivity in manufacturing grew 3.1 percent a year faster in Japan than in the U.S. between 1973 and 1986 and these trends continue today. Economists cite a number of factors that contribute to this decline:

- Shifts from a manufacturing-dominated to service-dominated economy
- Inadequate levels of investments in new plant, new equipment, and R&D to support innovation
- Decline in educational quality and in the number of graduate engineers and scientists
- Deterioration in work effort or a reduction in the traditional U.S. work ethic
- Government regulations
- Management failures

The electric utility industry plays a crucial role in curtailing the drop in economic efficiency. Since electricity provides the key energy supply in the modern economic chain, electric utilities can improve the U.S. economy's productive capacity by improving economic efficiency. Without continuing improvements in economic efficiency, the country cannot sustain economic growth and society will suffer stagnant or falling living standards.

No longer can utilities plan and operate power systems inefficiently or allow their customers to use electricity inefficiently. In addition, they must devise innovative technologies. A recent study concludes that, "... the two key innovative technologies for electricity generation that were developed in the 1960s (inexpensive coal and nuclear power plants) failed to live up to expectations in the 1970s...this failure...tracks the poor productivity performance of the industry."(6-8) Around 1968, the economies of scale for new electric production stopped improving. Industry productivity levels could no longer support inefficient power system planning, operation, and use.

A Look at Future Energy Use and Efficiency Issues

Case Study: U.S. versus West Germany

Comparisons among energy consumption patterns in the U.S. and other countries yield a different perspective of "where we are" and "where we want to go" in terms of energy efficiency and policy issues. In one 1975 study, researchers from the then Federal Energy Administration (FEA) examined the differences in per capita energy consumption between the U.S. and foreign countries. Researchers used West Germany as the basis for comparison. In 1990, the conclusions were still valid. (6-9) (Obviously, the recent reunification of East and West Germany will require alteration of these conclusions in future studies, given East Germany's vastly outdated infrastructure.)

The 1975 study examined and explained the differences in per capita energy consumption between the U.S. and West Germany; researchers wanted to quantify the factors involved. During the time period examined (just prior to 1972), West Germany used only about *half as much* energy per capita as the U.S. and although both countries have become more efficient, the ratio remained roughly similar through time.

As indicated in Table 6-1, the 1972 per capita energy consumption in West Germany was substantially lower in each of the sectors than in the U.S. While social and cultural differences contributed to this disparity, West Germany's use of newer, more energy efficient technologies also accounted for much of the difference.

Table 6-1

Sector	Per Capita Energy Consumption in West Germany as Percent of that in the United States—1972
Household and Commercial Industry Transportation	52.1 % 60.8% 25.7%
All Sectors	48.6%

In West Germany, the per capita energy consumption during the study period was much lower for all transportation modes except passenger rail travel. Road transport was the principal energy user in both countries, accounting for 74 percent of the U.S. transportation energy use and 79 percent in West Germany. The two principal factors accounting for the much lower per capita use of energy for transportation in West Germany were: 1) lower amounts of both passenger travel and freight transport, and 2) lower fuel consumption rates per vehicle-mile for passenger cars.

When corrected for its overall colder climatic conditions, West Germany's per capita energy use for space heating was only about one-half that in the U.S. Reasons for this difference included the larger size of U.S. dwellings, a greater percentage of dwelling space heated, a smaller percentage of dwelling units in apartment buildings, and lower use of insulated windows. The average floor space of dwellings in the U.S. was 57 percent greater than in West Germany. Only 45 percent of the dwelling space in West Germany was heated, and bedrooms were often left unheated even when heating was available. Although adoption of this practice in the U.S. would be difficult to implement, improvements in residential heating systems (such as greater temperature control in different parts of the house) could address this issue and result in energy savings. In addition, although the FEA study disagrees, most U.S. research indicates that West Germans used lower temperature settings on cold days than U.S. consumers.

Excluding space heating, the study found that West German residential energy use for other purposes was only one-fourth that of Americans. The percentages of households with the various electrical appliances were not substantially different, except for air conditioners and clothes dryers, which were found in negligible numbers in West Germany. Dishwashers were present in much smaller percentages of households in West Germany and refrigerator sizes were much smaller.

In West Germany, the per capita use of energy for hot water was only about one-third that of the U.S. This comparison with West Germany suggests that the U.S. may be able to achieve far greater potential energy savings than previously considered. West German homes and businesses commonly install small point-of-use systems that heat water only when needed. With these systems, the user becomes more closely involved with the system's activation and operation. The adoption of such systems in the U.S. might alter hot water consumption practices.

During the period the study, U.S. industry used 40 percent more energy than West Germany, as Table 6-2 shows. U.S. energy use in relation to output was also higher when output is measured in physical rather than monetary terms. The lower per capita industrial energy consumption in West Germany is due in part to the lower per capita industrial output and in larger part to the lower levels of energy used in relation to output.

Table 6-2

Comparison of Industrial Energy Consumption in
Relation to vValue of Shipments for the United States
and West Germany—1972

Industrial Sector	10^3 Btu/$ of Shipments		West Germany as Percent of United States
	United States	West Germany	
Food	11.9	8.3	70
Paper	104.0	38.6	37
Chemicals	71.8	40.8	57
Petroleum and coal products	112.0	56.0	50
Stone, clay, glass, and concrete products	75.3	54.8	73
Primary metals	97.0	77.6	80
Total for six energy-intensive industries	61.2	42.2	69
Other manufacturing	9.4	7.1	76
Industry total	34.8	25.1	72

Energy consumption in West Germany's six energy intensive industries ranged from 20 to 63 percent less, in relation to value of product, than the same industries in the U.S. For example, energy use per ton of petroleum products, steel, and paper was 33, 32, and 43 percent, respectively, less than in the U.S.

West German industry's lower energy use may be due, in part, to the large amount of electricity (28 percent) generated by self-producers, principally industrial establishments. This combination of electricity and process heat permits greater thermal efficiency.

Just prior to German reunification, these statistics had changed slightly—but not dramatically. This large disparity in energy use between the two countries suggests that U.S. economic growth and standard of living improvements could continue without a proportionate increase in energy consumption. The U.S. would only need to make energy efficiency an actual priority, instead of simply pay lip service to it. It would mean adopting a whole new generation of energy efficient appliances (and developing even better ones) as well as reorganizing certain traditional industrial processes. This step would lead the U.S. closer to parity with West Germany. Compromises in "quality of life" issues would move the U.S. even closer. However, past experiences show that most Americans are not willing to make these choices.

Future Efficiency Improvements Due To U.S. Government Mandates

Researchers estimate that efficiency improvements attributable to mandated U.S. government standards and market forces will reduce annual energy use by 8.5 percent in the next eight years (Figure 6-6). This reduction represents an aggregate savings of approximately 323,000 GWh. These improvements include changing energy prices, economic activity, and the normal penetration of efficient technologies.

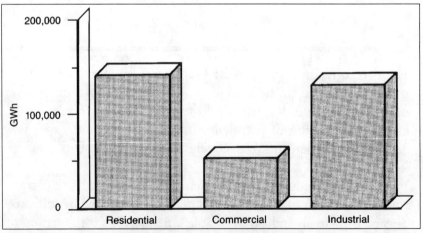

Figure 6-6. Estimates of efficiency improvements attributable to regulatory mandates and natural market mechanisms in the year 2000—by sector.
* Versus load growth with efficiency levels frozen

In developing these estimates, researchers first established a baseline forecast of end-use consumption by combining information from three main sources: aggregate energy projections from the North American Electric Reliability Council, end-use projections from several EPRI end-use models, and equipment efficiency trends from a variety of industry sources. Next, they developed a constant efficiency scenario of future energy consumption by freezing the efficiency levels of post–1987 equipment at the efficiency levels of the 1987 stock. Finally, they estimated the efficiency savings as the difference in energy consumption between the two scenarios.

Future Efficiency Improvements From Utility DSM Programs

Utility DSM programs will produce substantial reductions in annual energy use and summer peak demand in future years. For example, in the year 2000, analysts expect that utility DSM programs will reduce annual energy use by as much as 6 percent, and summer peak demand by 6.7 percent (Figures 6-7 and 6-8). This demand reduction is equal to forty-five gigawatts and total capi-

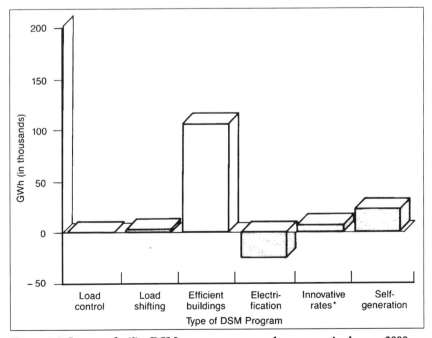

Figure 6-7. Impact of utility DSM programs on annual energy use in the year 2000.

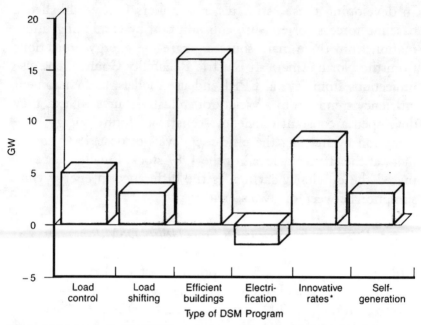

Figure 6-8. Impact of utility DSM programs on summer peak demand in the year 2000.

tal savings of $45 billion. (Comparable estimates of reductions in energy use and summer peak demand attributable to utility DSM in the year 2010 are 5.7 percent and 9.6 percent, respectively.) (6-10) When these estimates are combined with naturally occurring end-use efficiency increases, researchers project even greater reductions. In the year 2000, projections suggest an 8.5 percent drop in electric energy use; thus, the combination of naturally occurring efficiency improvements and DSM will yield a drop of almost 15 percent.(6-11)

Chapter 7 links the topics of electricity, energy efficiency, and economics with issues such as technology and productivity, and explains the role the U.S. electric utility industry can play in the future.

References:

6-1 A. Fickett, C. Gellings, and A. Lovins, "Efficient Use of Electricity," *Scientific American*, September 1990.

6-2 *Utility Energy Strategies: The Role of Efficiency, Productivity, and Conservation.* EPRI, CU-6272, February 1989.

6-3 The Gallup Organization, Inc., for the National League of Women Voters, December, 1987.

6-4 Emerging Issues Facing American Utilities and the Nation, Jeff Banks, Cambridge Reports/Research International, APPA Conference, June 17, 1992.

6-5 "New Push for Energy Efficiency." *EPRI Journal*, April/May 1990.

6-6 Yau, T. S. "Greenhouse Gases—The Role of Electric End Use." EPRI, June 1989.

6-7 *Electric Van and Gasoline Van Emissions: A Comparison.* EPRI, TB.CU.177, October 1989.

6-8 Energetics, Inc., *Utility Energy Strategies: The Role of Efficiency, Productivity, and Conservation.* Palo Alto, Calif.: Electric Power Research Institute, CU-6272, February 1989.

6-9 R. L. Goen, et. al., *Comparison of Energy Consumption Between West Germany and the United States*, Federal Energy Administration, June 1975, Report PB-245 652.

6-10 *Estimating Efficiency Improvements Embedded in Electric Utility Forecasts.* EPRI, CU–6925, August 1990.

6-11 *Impact of Demand-Side Management on Future Customer Electricity Demand: An Update.* EPRI, CU-6953, October 1990.

CHAPTER 7
Technology's Impact on Energy Efficiency

Elements such as innovation and technological change affect energy and economic efficiency as well as industrial productivity. Table 7-1 describes the relative importance of technological change to economic efficiency, growth, and development. It compares the major elements of the traditional economic process with the modern economic process. The remainder of this chapter briefly discusses the importance of technological developments in the residential and commercial sectors. It then discusses the industrial sector in greater detail.

The Importance of New Technology Developments

Analysts estimate the maximum technical potential (MTP) improvement in end-use energy efficiency at between 8000 and 15,000 trillion Btu (eight to fifteen quads). (Refer to Table 7-1 and Figure 7-1.) This figure represents the technically possible efficiency gain that could result if the most efficient electric technologies known today were to attain complete market saturation in the year 2000. Analysts attribute approximately 96 percent of this estimate to upgrading electrical equipment and approximately 4 percent to electrification. (7-1)

Overall, the U.S. uses only 7 percent more energy than it did in 1973, yet the GNP has increased some 46 percent. Efficient use of electricity alone has already saved the U.S. at least $21 billion by delaying the need for new power plants. The advanced electric end-use technologies available today could help the country save even more on its energy bill. If today's most efficient elec-

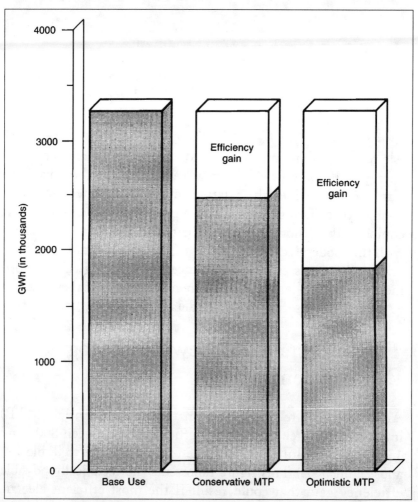

Figure 7-1. Technically possible efficiency gain in the year 2000 attributable to upgrading electrical equipment.

tric end-use technologies were applied in every possible case, they would have the potential to save the U.S. anywhere between 25 percent and 50 percent of the electricity it will use in the year 2000. The low-end estimate alone, which translates into 800 billion kWh, is enough to meet the entire energy needs of the eleven western states in the year 2000. Table 7-2 details these estimates. The following sections summarize the tables.

Table 7-1
Base Case Usage and MTP Impacts of Electricity-Saving Technologies

End Uses	1987 Base GWH	2000 Base GWH	Range of MTP Impacts Year 2000 (GWH)
Residential:			
Space Heating	159,824	223,024	71,915 - 122,285
Water Heating	103,499	134,509	43,481 - 88,995
Central A/C	78,127	90,134	26,265 - 30,996
Room A/C	15,254	13,063	2,421 - 4,222
Dishwashers	15,308	23,707	1,240 - 6,233
Cooking	30,390	39,271	3,115 - 7,132
Refrigeration	146,572	139,255	30,716 - 66,896
Freezer	59,779	48,073	11,534 - 15,594
Residual Appliances	240,861	353,620	98,242 - 141,552
Total Residential	849,614	1,064,656	288,929 - 483,905
Industrial:			
Motor Drives	570,934	780,422	222,226 - 351,040
Electrolytics	98,193	138,273	25,950 - 41,124
Process Heating	83,008	125,274	9,928 - 16,606
Lighting	84,527	114,097	19,016 - 38,032
Other	8,453	9,192	0 - 0
Total Industrial*	845,266	1,167,413	277,119 - 446,802
Commercial:			
Heating	77,245	128,322	16,335 - 30,333
Cooling	154,299	208,106	62,432 - 145,674
Ventilation	76,959	96,094	28,828 - 48,047
Water Heating	24,068	39,794	15,917 - 23,876
Cooking	16,172	26,381	5,276 - 7,914
Refrigeration	60,883	81,652	9,925 - 27,857
Lighting	238,488	283,124	62,916 - 157,291
Miscellaneous	108,447	177,254	32,228 - 64,456
Total Commercial	756,561	1,040,727	233,857 - 505,448
TOTAL	2,451,441	3,272,796	799,905 - 1,436,155

* Sum of end uses may not add to total due to rounding

Table 7-2

Technically Possible Efficiency Gain in the Year 2000 Attributable to Upgrading Electrical Equipment and Electrification

	Trillion Btu	
	Conservative MTP	Optimistic MTP
Upgrading Electrical Equipment[1]	7,999	14,361
Electrification[2]	305	548
	8,304	14,909

[1] In GWh=799,905 to 1,436,154   [2] Net of generation losses

The gap between the high and low estimates is indicative of how difficult it is to determine the efficiency potential of technologies that have not yet been widely deployed and thoroughly tested in the marketplace. New equipment may perform quite differently under various real world conditions and use patterns.

Many of today's most efficient electric technologies are the focus of EPRI research, development, and demonstration projects on behalf of the U.S. electric utility industry. In addition, EPRI researchers want to identify promising emerging technologies and to improve the attributes of existing technologies—to reduce their size and cost, extend their lifetimes, and attain even greater efficiency levels.

Residential Sector Energy Efficiency Savings

In the residential sector, energy efficiency measures could save between 27 percent and 46 percent of the electricity projected for use in the year 2000 (between 289 billion and 484 billion kWh). In this sector, the greatest savings could come from technological advancements in water and space heating, lighting (which represents 45 percent of residential energy use totals), and miscellaneous appliances. To achieve this level of efficiency, homeowners would weatherstrip and caulk homes, install storm windows and doors, and add more insulation to ceilings and floors. They would purchase and install efficient electric heat pumps and solar panels to reduce energy use for space and water heating. Finally, they would use compact fluorescent bulbs with incandescent-like color spectrums in all light fixtures.

The newest models of the electric heat pump yield impressive energy savings, when compared with conventional models. EPRI, in collaboration with the Carrier Corp., the country's leading air conditioner manufacturer, developed a heat pump that is 30 percent more efficient than older models. Introduced to the market in 1989, this heat pump represents the most efficient heating, ventilating, and air conditioning appliance available today. In the summer, it can recover heat rejected from its air conditioning function and use it to heat water: the result is "free" hot water for the customer. An optional control can automatically limit the amount of electricity the heat pump uses during a utility's hours of peak demand. Benefits to the customer include reduced energy costs, extended longevity, and greater comfort—no "cold blow" or noise problems.

Recent projects have attempted to quantify the potential to improve energy efficiency among residential customers. For example, in 1983, the Hood River Conservation Project, a $20 million, five-year experiment, launched a residential weatherization program that targeted electrically heated homes in Hood River County, Oregon, a community of 15,000. The Bonneville Power Administration funded the program and Pacific Power & Light managed it.

To encourage customer participation, Bonneville paid for the weatherization measures and offered free energy audits to every eligible home in the county. Under the auspices of the program, several measures, including ceiling and floor insulation, storm windows, caulking, and weatherstripping were installed in about 3000 homes. This program did not replace water- or space-heating equipment. The project drew 85 percent participation from eligible homeowners and resulted in annual savings averaging 2600 kWh per house.

While participation in the project was healthy, it wasn't automatic. When necessary, project workers went door-to-door to sign up residents. Some residents refused to participate simply because they did not want project workers in their homes. Others were in the process of selling their houses and did not want to be bothered. Before the launching the project, BPA hired a sociologist to study the community and determine beforehand how customers would react. As a result, the program's marketing message was structured so that "proud" residents would not feel they were getting a handout. Its basic message: the community could set an example for the rest of the country.

Commercial Sector Energy Efficiency Savings

In the commercial sector, energy efficiency improvements could save between 23 percent and 49 percent of the electricity demand in the year 2000 (between 234 billion and 505 billion kWh). The biggest contributors to these potential savings include lighting, space cooling, and various "plug loads," such as office equipment, computers, and copy machines.

To achieve such savings, typical commercial buildings—office towers, restaurants, or retail stores—would integrate a variety of energy efficient technologies into their daily operations. Sophisticated lighting systems would include more efficient bulbs and ballasts and a variety of controls (such as occupancy sensors that turn lights on and off automatically and computer-based timers). Another type of control, based on photocell technology, can track the amount of daylight entering a room and trigger an appropriate supplement from the electric lighting system when necessary. Building windows would be specially treated to allow visible light through while keeping undesired heat out. High efficiency motors would power elevators and escalators.

Results from an ongoing EPRI project indicate that many commercial buildings can save electricity by incorporating a lighting system design that integrates with the heating, ventilating, and air conditioning systems. Findings show that if only 5 percent of existing U.S. commercial spaces and 25 percent of new construction adopted more efficient strategies each year, they could save 1400 MW annually, the equivalent output of one and a half large power plants. These better designs would dramatically reduce peak loads for utilities, particularly during the summer, when air conditioning systems create substantial demand in many parts of the country.

Industrial Sector Energy Efficiency Savings

The industrial sector could save between 24 percent and 38 percent of energy used in the year 2000 (from 277 to 447 billion kWh). In typical industrial applications, electricity-based processes are at least twice as efficient as their fossil fuel alternatives. This rating accounts for the 65 percent of energy typically lost when converting "primary" energy into electricity. Often, fossil fuel energy is wasted from heating things other than the product material—ambient air or a container, for example. Electricity offers better control, allowing users to focus energy exactly where they need it.

The biggest opportunity for energy efficiency improvement in this sector lies in more efficient motor drives, which account for 67 percent of the industrial electricity used today. Essentially, motors keep American industry moving—they turn its fans and blowers, run its pumps and compressors, spin its blenders, and propel its conveyors and process lines.

To take advantage of the full potential for savings, industries would begin using high efficiency motors with adjustable speed drives (ASDs), which allow for efficient and smooth variations in motor speed. Work areas would be illuminated with efficient lighting systems that incorporate high-frequency ballasts, improved reflective fixtures, and sophisticated lighting controls like those found in the commercial sector.

Electronic ASDs use semiconductor and switching circuits to vary the frequency of power to alternative current motors.

Without them, these motors operate at a constant speed that only mechanical and hydraulic devices like clutches, gears, and valves regulate. ASDs give motors only as much power as they need to accomplish the task at hand. They also operate much more quietly than do mechanical and hydraulic devices, and they allow for smooth startups and shutdowns, which reduces equipment wear. EPRI studies have shown that electronic ASDs alone can slice an industry's energy use requirements by 20 percent. With ASDs, U.S. industry could save 95 billion kWh of energy per year, or about $5 billion annually.

Factors Affecting Industrial Productivity

Over the long term, industrial productivity depends on overhauling out-moded operations and introducing new technologies. Industry has made some improvements, but they only represent first steps. In its recent study, *Made in America*, the MIT Commission on Industrial Productivity reports: "In such areas as product quality, service to customers, and speed of product development, American manufacturers are no longer perceived as world leaders, even by American consumers. There is also evidence that technological innovations are being incorporated into practice more quickly abroad, and the pace of invention and discovery in the United States may be slowing."

More than two-thirds of U.S. manufacturing output now faces direct foreign competition. If manufacturers want to make the cost and quality of American goods more attractive consumers, then they must adopt new ways to incorporate the latest technological advances into both factories and products. The current effort to revive industrial productivity is based on two key assumptions: that the American economy still depends on manufacturing and that manufacturers must base long-term productivity gains on technological innovation rather than on cost-cutting expedients.

The first of these assumptions stirs considerable debate, since manufacturing's contribution to the GNP has declined to less than one-fourth and its share of workers to less than one-fifth.

In the book *Manufacturing Matters: The Myth of a Post-Industrial Economy*, Stephen S. Cohen and John Zysman of the University of California at Berkeley argue that a "direct linkage" exists between many service jobs and manufacturing production. Although immediate employment in manufacturing accounts for only about 21 million jobs, the authors estimate that two to three times as many Americans—most considered "service workers"—depend directly on manufacturing for their livelihood. "If manufacturing goes, these service jobs will go with it," they conclude.

Table 7-3
Electricity Plays a Key Role in the Technological Change Process

Elements	Traditional Economic Process Chain	Modern Economic Process Chain
Factors of Production	Labor Capital Materials Energy	Labor Information Technologies Financing Materials Energy
Technology Base	Assembly Line Mechanization Mass Production Internal Combustion Engines	Computerization Telecommunications Automation Electronics Rapid Transportation
Energy Requirements	Process Steam Oil and Coal for Industry and Transportation Low Cost Energy	Electricity High Efficiency Targeted Applications Environmentally Sound
Process Outcomes	Manufacturing-Based Economy Economies-of-scale Urbanization Environmental Quality Ignored	Service-oriented Economy International Marketplace Information Industry Environmental Quality Important

The MIT commission agrees, directly attacking the proposition that a transition from manufacturing to services produces an inevitable and desirable stage in economic development. "We think this idea is mistaken. A large continental economy like the United States will not be able to function primarily as a producer of services in the foreseeable future," the commission's report states bluntly. "The United States thus has no choice but to continue competing in the world market for manufacturers."

The ability to compete depends, of course, on a variety of factors. Manufacturers can directly control several factors that affect productivity—including shutdown of obsolete facilities, retrenchment to the most profitable niches of a market, and introduction of new technology. In today's increasingly technical world, cost cutting without innovation amounts to retreat; long-term survival depends on finding new ways to develop and use technology.

Adam Kahane[7-3] recently published a series of three case studies of the iron and steel, pulp and paper, and cement industries (see Figures 7-2, 7-3 and 7-4). These studies compare energy use in France, the U.S., Japan, and Sweden. Kahane's analysis reveals the trend toward electrification and toward efficiency.

In the iron and steel industry, all countries but Sweden saw the consumption per ton of raw steel rise between 1973 and 1984 as manufacturers deployed new technologies. In the pulp and paper industry, electricity consumption per ton of product decreased in France and Japan between the mid 1970s and 1984 and increased in Sweden and the U.S. Concurrently, electricity use dropped in France, Japan, and Sweden.

In the cement industry, Swedish and U.S. unit energy consumption rose between 1972 and 1984 and fell in Japan. The increase in Swedish and U.S. unit electricity consumption resulted from the replacement of the wet process with the more electricity intensive dry process. The Japanese made the same process decisions while employing even more efficient plant designs and further reducing energy consumption.

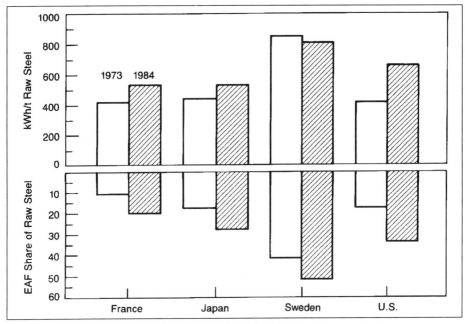

Figure 7-2. Steel industry electricity use, kWh/t. Calculated from total electricity consumption, purchased plus self-generated, and metric tons of raw steel produced. Source: (7-3)

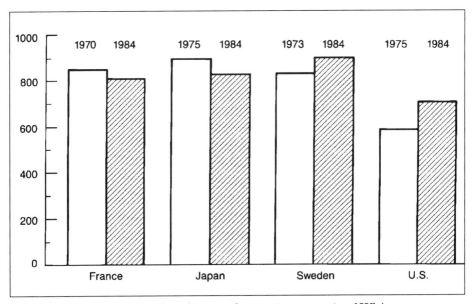

Figure 7-3. Changes in pulp and paper industry unit consumption, kWh/t. Calculated from total electricity consumption, purchased plus self-generated, and metric tons of pulp and paper produced. Source: (7-3)

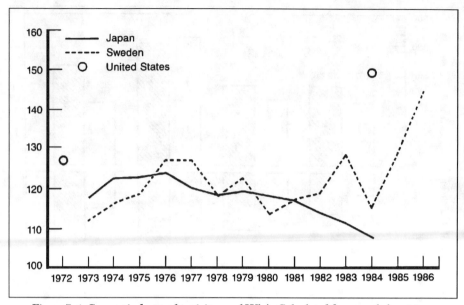

Figure 7-4. Cement industry electricity use, kWh/t. Calculated from total electricity consumption, purchased plus self-generated, and metric tons of cement.
Source: (7-3)

In reviewing the implications and conclusions of these studies, Kahane summarized: "Many of the new technologies installed in response to external pressures use electricity. The most common examples are technologies for automation, pollution control, water, waste and heat recovery, additional product finishing, and specialized heating. In general terms, electricity contributes to the solution of many industry 'problems'."

"If all other aspects of production were unchanged, then, we could expect that technological change would generally push up unit electricity consumption. The reason why unit consumption actually decreased in some cases (after correcting, as far as possible, for changes due to product mix) is that *other aspects of production did change*, in response to these same or other external pressures."

Trends in Industrial Productivity: A Situation Analysis

After a long period of declining productivity, U.S. industry shows signs of resharpening its competitive edge. The improvement has been particularly impressive in the manufacturing sector, where productivity has been growing faster over recent years than in most all other major industrialized nations. U.S. manufacturing productivity, in terms of output per unit of combined labor and capital input, increased fivefold in recent years—from an average of about 0.5 percent during the 1973–1979 period to about 2.5 percent during the 1979–1986 period. (7-1)

This much-needed recovery may be very fragile. Before the 1990 recession, factories operated at near full capacity, however, more than a third of the growth in labor during this period came at the expense of lost jobs. Rising exports helped the U.S. trade deficit, but much of the improvement resulted from depreciation of the American dollar.

Over the long term, industrial productivity depends more on overhauling outmoded operations and introducing new technologies than on one-time belt-tightening. Electric utilities can play a major role in this effort by helping their industrial customers improve productivity through the increased use of highly efficient electric technologies.

Current efforts to revive industrial productivity are based on two key assumptions. First, manufacturing is still important to the American economy. After a period of deemphasis during the Reagan administration, economists and planners are recognizing the important link between service jobs and manufacturing production. In *Made in America*, the MIT Commission attacked the once popular proposition that a transition from manufacturing to services is an inevitable and desirable stage in economic development. Instead, the Commission strongly affirms that the U.S. will not (and cannot) function solely as a producer of services; thus, the country must continue to compete in the world market for manufacturing jobs.

Second, long-term productivity gains must be based on technological innovation rather than on inherently limited cost-cutting expedients. As technology rapidly evolves, those industries that fail to innovate will simply concede defeat in the marketplace. Industries that seek to upgrade their manufacturing processes will succeed with the help of electrotechnologies. Electricity's unique advantages in automation and precise delivery of energy offer numerous benefits, including shorter production times, improved product quality, reduced energy consumption, and fewer environmental impacts.

The Importance of New Technology and Productivity Increases: A Lesson From the Steel Industry

Recent experience in the steel industry illustrates vividly how American companies are increasing their productivity levels. Faced with increased competition from overseas and a steady loss of market share, U.S. steel companies went through a wrenching transformation between the mid–1970s and mid–1980s. As steelmakers cut costs, closed inefficient plants, and reduced overall capacity by about 28 percent, the industry lost an estimated two-thirds of its jobs.

As a results of these drastic changes, productivity soared. The average cost of a ton of steel in the U.S. fell from $550 to $490, while Japanese costs have been rising. U. S. steel exports nearly doubled last year and may double again this year, according to Elizabeth Bossong, formerly manager of economic research at USX Corp. The average labor required to make a ton of steel from ore in an integrated mill dropped from about ten man-hours in 1982 to less than six man-hours now, she says, with some individual plants doing considerably better.

In addition to the dramatic cost cutting and the declining dollar, Bossong cites *the adoption of new technology* as an important factor in reviving productivity. Specifically, continuous casting has largely replaced the older practice of casting and then reheating separate ingots. A decade ago, continuous casting produced only about 16 percent of raw steel output; now it yields more than 60 percent.

"Since 1973 [U.S. economic] growth has slowed about 2.8 percent, and productivity has dropped by almost two-thirds, to only 1.1 percent," Bossong recently told the Steel Survival Strategies Forum in New York. "What we need desperately in the country is to get those productivity numbers growing once again at something like 2.9 percent—certainly back to 2 percent—if we're going to achieve the kind of living standard potential what we used to assume was our birthright." (7-4)

The importance of using technology to improve productivity can be seen in other industries as well. The chemical and aircraft industries, for example, both remain world leaders in technology and enjoy a healthy trade surplus. To maintain their technological advantage, both have recognized the value of collaborative research. Leading chemical companies have established joint R&D programs at several universities, and the Aerospace Industries Association coordinates research by private companies, the government, and universities. This research seeks to develop eight key technologies needed for the 1990s.

By contrast, technological weakness continues to devastate the machine tool and consumer electronics industries. As machine tools made a rapid transition to electronic controls, American manufacturers fell behind their Japanese and West German counterparts. Although many of the technologies involved in consumer electronics were originally developed in the U.S., foreign companies have applied them more aggressively. Both industries have now established collaborative R&D efforts to help them catch up, but the programs remain relatively small.

The Electric Utility as a Technology/Productivity Bridge for Industrial Consumers

Electric utilities can play a major role in improving U.S. manufacturers' productivity levels by helping their industrial customers adopt new electrotechnologies. Many of the new technologies that can boost American industry's productivity use electricity because of its unique advantages in automation and precise delivery of energy. Electrotechnologies allow manufacturers greater flexibility to alter process time and temperature. They can shorten production times, improve product quality, reduce energy consumption by heating products more exactly, and often curtail environmental problems at a factory as well.

Steelmaking again provides several good examples of how electricity can increase productivity by fundamentally changing time and temperature components. Arc furnaces provide an intense concentration of heat to produce steel quickly in controlled quantities from recycled scrap. The result: "mini-mills" increasingly compete directly with large, integrated mills for the high end of the steel market. Continuous casting uses electrically-driven and controlled equipment to eliminate reheating steps, which consume both time and energy. High-power lasers and electron beams, which deliver energy with unprecedented precision, have created new applications in welding. And very high temperature plasmas, ionized gases that can conduct an electric arc, may eventually provide a way to produce iron from ore directly, without the need for coke or a blast furnace.

Two hurdles, however, impede the introduction of advanced technologies into wider industrial practice. The first of these results from America's declining commitment to industrial research and development relative to its international competitors. Industry-funded R&D in this country has leveled off at about 1.3 percent of gross domestic product; in comparison, the figures for West Germany (prior to German reunification) and

Japan are about 1.6 percent and nearly 2 percent, respectively, and both of these percentages continue to grow rapidly. At the same time, government-sponsored R&D in the U.S. remains largely defense industry related and has become less likely to contribute new technologies for industry.

"Military research and development no longer have the spin-off effect on the civilian sector of the economy that they did in the period when computers and microelectronics were developing," according to a recent Brookings Institution study, *Restructuring American Foreign Policy*, sponsored by Carnegie Corp. of New York. "U.S. businessmen now increasingly complain that American innovations in basic science and technology are being commercialized abroad," the study concludes. "These industrial rivals, moreover, have come to undertake their own technical investments in amounts comparable to the U.S. investment, and since they direct a higher proportion of it to commercial [rather than military] purposes, their investments now exceed our own in that area."

The second hurdle involves the difficulty of transferring existing technologies to industries that could use them most effectively. This problem is particularly acute for the roughly quarter million small and medium-sized manufacturers in the U.S.; these businesses generally lack both R&D capabilities of their own and even an established procedure for surveying and adopting technological innovations. At a time when many of their competitors around the world actively seek out the best available technologies and put them to use in more flexible production systems, many smaller American companies need help just to catch up.

Electric utilities sit in the ideal position to provide some of this help. For example, small firms that need new cutting tools do not know where to go for help. They may face decisions among different options (such as a laser, a water jet, and a saw) or experience difficulties when integrating a new piece of equipment into their production line. For example, electric utilities could step in and help these customers by offering an exciting new service option that responds to customer needs for data and information on new electrotechnologies.

An electric utility with a strong industrial program will improve both a customer's and its own competitive positions: the industrial customer modernizes and improves its overall efficiency and profitability and the utility retains this customer's load.

References:

7-1 *Efficient Electricity Use: Estimates of Maximum Energy Savings*. EPRI, CU-6746, March 1990.

7-2 "The Return of American Industry," *EPRI Journal*, October/ November, 1989.

7-3 *Technology Change and Industrial Electricity Use: Towards Better Generalizations*, Kahane, A., Shell International Petroleum Co., London UK

7-4 *Electricity in Economic Growth*, National Academy Press, Washington, D.C., 1986.

SECTION III.

INTRODUCTION:

Electricity and Industry

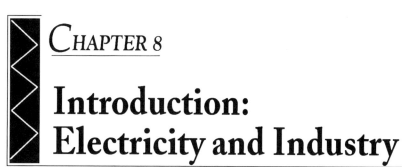

CHAPTER 8
Introduction: Electricity and Industry

Electricity's impact on U.S. industry has increased steadily since the Industrial Revolution. During the past thirty-five years, in particular, U.S. industry has become less energy intensive but more electricity intensive. In fact, the reduction in U.S. total energy use, and particularly in fossil fuel use, is due in large part to our efforts to electrify—use more electricity and less total energy (see Figure 8-1). This chapter briefly examines the history of industrial electricity use and then discusses the present and future impacts of electricity on industry.

Industrial Electricity Use: A Historical Perspective

Early stages of the Industrial Revolution relied on energy supplied from the waterwheel and the windmill. However, waterwheels and windmills posed two major limitations: they could capture only a small amount of the power available from falling water and wind, and they were fixed to specific sites which were limited in number. Until the 1870s, water provided the primary source of motive power. In fact, until 1900, the Census of Manufacturers listed "access to water" as one of seven key factors contributing to siting choices.

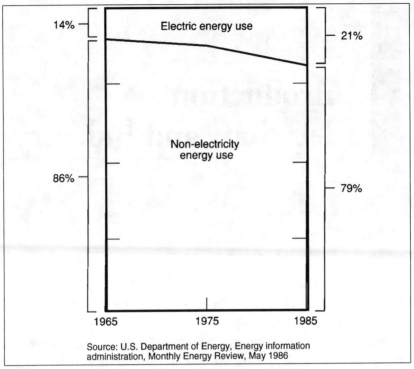

Figure 8-1. Electricity's share of energy use in manufacturing.

During the 1870s, steam power began replacing the waterwheel and the windmill as U.S. industry's predominant energy source. By 1900, industrial businesses preferred steam power over water power by a ratio of six to one. Steam-generated power freed manufacturing from the locational and sometimes seasonal constraints imposed by falling water because, theoretically, steam-supplied factories could operate anywhere in the country.

However, another factor—practical economics—often dictated their siting locations. Because of the high costs of transporting coal, it made sense to locate close to a coal source. In addition, until the development of effective transmission and distribution systems, industries had to use a steam engine's mechanical energy at the same location it was generated. Moreover, because steam engines were inefficient below a certain size, industries developed large manufacturing sites and concentrated them in the urban core of a limited number of cities.

Electricity eclipsed steam as a primary industrial power source at the beginning of the twentieth century. Electric motors reduced the cost of driving machinery because of their greater efficiency. In addition, they offered flexibility (in space, time, and scale), which allowed manufacturers to innovate their factory configurations as well as management structures. Finally, and perhaps most important to individual states in the U.S., the evolution of transmission and distribution systems allowed electricity generated at a central site to be transmitted to distant work sites.

Manufacturers could now make siting decisions based on other practical and economic considerations. For example, other siting variables have become increasingly important to industrial businesses, including:

1. Proximity to raw materials
2. Proximity to markets (transportation costs, highway/rail networks)
3. Availability and cost of labor (skilled and unskilled)
4. Availability of capital
5. Capitalistic enterprise and management
6. Climate
7. Government inducements and taxation

After World War II, an enormous expansion of electric generating capacity occurred. In 1950, there were only 8,200 miles of transmission lines rated at or above 230 kilovolts. By 1980, there were over 100,000 miles of such lines.

Today, approximately thirty formal or informal power pools coordinate operational plans, and nine regional reliability councils work together to manage system planning and reliability. Industries in all regions of the country have access to electricity grids, each handling enormous amounts of power.

Electric T&D Systems Spur Industrial Regionalization

As industry began spreading throughout the U.S., previously undeveloped states added manufacturing businesses, which enhanced the growth of these local economies. As of 1950, roughly three-fourths of all manufacturing employment still existed in the traditional manufacturing centers of the northeastern industrial cities. After 1950, however, manufacturers began shifting their operations to other regions of the U.S.

The share of manufacturing employment in the South and West rose from 26 percent in 1947 to 43 percent in 1977. Moreover, in 1977, manufacturing employment in the South exceeded manufacturing employment in the Northeast. By 1977, the traditional manufacturing cities of the northeastern U.S. contained no more than 58 percent of all factory workers. Some historians claim that this decentralization of U.S. industry produced a stronger, more robust democracy. This affinity for technological determinism was not confined to the North American continent: Lenin was reported to have claimed "communism is socialism plus electricity."

For perspective, the remainder of this section briefly examines regionalization's impacts on three separate industries: textile, automotive, and steel. Electricity plays an important role in each of these industries and electrotechnologies will help each industry meet the competitive challenges of today's global marketplace.

Textiles

Cotton textile mills were born in the Northeast where abundant water supplies powered the mills. By the 1860s, however, mills began using steam-engine-powered equipment, which freed them from their reliance on a water source. This ability to change energy sources started the migration of many mills from the Northeast to the South. Concurrently, industry owners discovered that they could pay unskilled southern laborers 20–30 percent less in wages and still achieve higher productivity levels. These additional economic benefits accrued due to the absence of labor unions, the existence of longer work hours, and the diligence of the southern worker. By 1930, only 30 percent of textile workers still labored in the Northeast (today the figure is roughly 7 percent).

Electricity also played a key role in this transition. In the early 1900s, the electric motor overtook all other energy sources—it could provide the industry with mechanical power much more cost-effectively. By eliminating the complex network of equipment needed to run a steam engine, the industry reduced the capital-intensive nature of its business—thus making it even more sensitive to labor cost differentials. In addition, siting restraints essentially disappeared. The raw material used by the textile industry (e.g., cotton) is easily and cost-effectively transported from almost any location. And last of all, the growth of electric transmission and distribution networks eliminated worries about power supplies.

Automotive industry

Since its inception, the automotive industry dispersed regionally at a much slower pace than the textile industry. The vast majority of the motor vehicle and equipment manufacturers located their businesses in the Central Northeast region of the country (centered, of course, around Henry Ford's home in Detroit) in the early 1900s. In 1929, fully 75 percent of all motor vehicle manufacturing employees were found in this region. As it developed, the Detroit area also drew the iron and steel industry because of the automotive industry's need for these products.

In more recent years, the percentage of motor vehicles manufactured in the Detroit area has declined steadily. This decentralization has occurred primarily in facilities which support the later stages of assembly, when electricity use is heaviest. Approximately 100 plants outside the Detroit area are now engaged in some phase of automobile assembly. Manufacturers have learned it is cheaper to transport unassembled vehicles to market areas and then assemble them, rather than ship totally assembled vehicles. In addition, there has been considerable concern over the worker–management conflicts in the Detroit area, including the high costs and low productivity of the workers. Another sign of regionalization comes from the recent decisions of domestic and foreign automobile producers to locate major production plants outside the Detroit area (in California, Tennessee, and Kentucky, for example). With the automotive industry now espousing the merits of the "just-in-time" assembly system based on low inventories and continual supplies, pockets of automotive suppliers will also begin locating near these new production plants.

Iron and steel industry

The iron and steel industry's innovative production techniques have eliminated many traditional siting factors. Historically, the iron and steel industry needed access to raw materials, particularly heavy iron ore and coking coal. Accordingly, the Upper Midwest and the Middle Atlantic regions, with their abundant coal and iron ore deposits, fulfilled this requirement.

Because of progressive advances in the industry's production techniques, its need for coal and iron ore diminishes each year. Electric arc furnaces present the most revolutionary development to date; these furnaces completely eliminate the need for iron ore and coal in the steelmaking process because they use scrap. Electric furnaces now produce nearly 40 percent of the raw steel output, a significant jump from the 10 percent share of the market it garnered in 1965.

As a result of this electric furnace technology, the industry now uses small-scale steelmaking operations, called mini-mills. These widely dispersed plants operate exclusively with electric furnace technology and serve local and regional markets. Mini-mills offer several key benefits: 1) relatively low investment costs; 2) half as many workers required per unit of output; and 3) only 33–40 percent of energy required per ton of steel produced. Given these advantages and the continued availability of scrap, the steel industry is likely to add more mini-mills and expand still further from its traditional geographic base.

Overview of Recent Industrial Electricity Demand

Since electricity for industrial use became generally available in the early 1900s, manufacturers have steadily increased their purchases. Concurrently, manufacturing fuel use has steadily decreased. These two phenomena—reduced fuel consumption and increased electricity use—point to an important trend often overlooked in energy analyses: U.S. industry continues to electrify its plants.(8-1)

Over the past decade, energy end-use patterns, in general, and industrial energy consumption patterns, in particular, have been

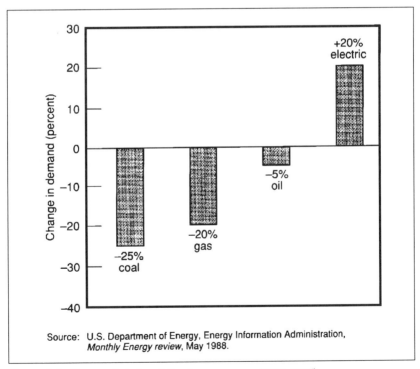

Source: U.S. Department of Energy, Energy Information Administration, Monthly Energy review, May 1988.

Figure 8-2. Changes in industrial end-use patterns (1975-1985).

in a state of flux. Figure 8-2, for example, shows the fuel-specific changes in industrial energy demand which occurred during the period 1975–1985. Higher wellhead prices help explain the downturn in industrial gas use and the stabilization in oil use reflects the worldwide surplus of residual fuel oil. However, the dramatic drop in coal use and the concurrent increase in electrification might prove surprising to some.

EPRI has grouped American industries into three categories, based on shared characteristics:

- Process Industries
- Metals Production
- Materials (metals and non-metals) Fabrication

The industry sectors in each category are listed by Standard Industrial Classification (SIC) code in Table 8-1. Because energy intensity (e.g., kWh per employee) is closely linked to industrial product and process, market segmentation by SIC code remains one of the most commonly used segmentation approaches in the industrial sector.

The process industries consume almost as much electricity as the metals production and materials fabrication industries combined. Process industries (including food, textile, pulp and paper, chemical, and petroleum) produce commodities in large plants primarily using continuous, around the clock processes. These industries produce consumer goods vital to the American economy (e.g., food, gasoline) as well as provide raw materials crucial to other industries (e.g. the PVC used in plastics).

Table 8-1
Industry Sectors by SIC Code

PROCESS INDUSTRIES	MATERIALS FABRICATION	
SIC 20 Food and Kindred Products	*METALS FABRICATION*	
SIC 21 Tobacco Products	SIC 34	Fabricated Metal Products
SIC 22 Textile Mill Products	SIC 35	Machinery, Except Electric
SIC 26 Paper and Allied Products	SIC 36	Electrical Equipment
SIC 28 Chemicals and Allied Products	SIC 37	Transportation
SIC 29 Petroleum and Coal Products	SIC 38	Instruments and Related Products
METALS PRODUCTION (SIC 33)	SIC 39	Miscellaneous Manufacturing Industries
FERROUS PRODUCTION		
SIC 331 Blast Furnace, Basic Steel Products	*NON-METALS FABRICATIONS*	
SIC 332 Iron and Steel Foundries	SIC 23	Apparel, Textile Products
SIC 339 Misc. Primary Metal Products	SIC 24	Lumber and Wood Products
	SIC 25	Furniture and Fixtures
NON-FERROUS PRODUCTION	SIC 27	Printing and Publishing
SIC 333 Primary Non-ferrous Metals	SIC 30	Rubber and Miscellaneous Plastics
SIC 334 Non-ferrous Rolling and Drawing	SIC 31	Leather, Leather Products
SIC 335 Non-ferrous Foundries	SIC 32	Stone, Clay and Glass Products
SIC 336 Secondary Non-ferrous Metals		

Source: *Industrial Electrotechnologies and Electrification—Volume 1: Electrotechnology Reference Guide*, EM-4527, developed by Resource Dynamics, Inc. for EPRI, April 1986.

Figure 8-3 disaggregates total energy use and electricity's share of this total for each industry grouping in 1984. The process industries accounted for 47 percent of industrial electricity use in 1984, with metals production and fabrication together using almost 40 percent. The seven sectors in the non-metals fabrication category accounted for roughly 14 percent of total electricity use.

Five specific manufacturing processes have the largest industrial energy requirements: paper, chemicals, petroleum, stone/clay/

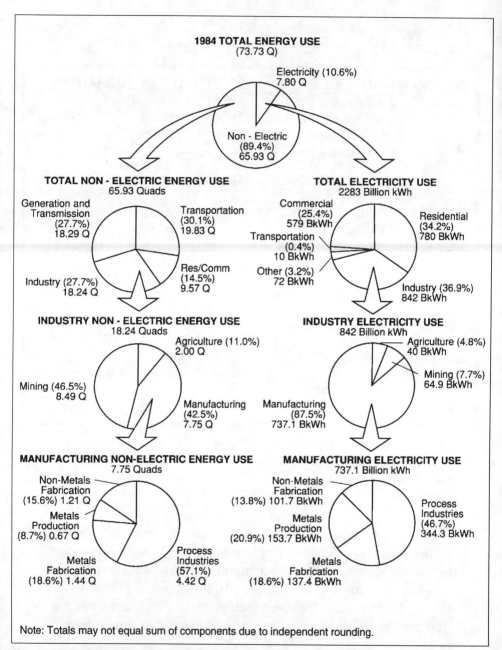

Figure 8-3. Energy and electricity use overview.
Sources: Annual Energy Review 1984, U.S. Department of Energy, April 1985; Federal Reserve Statistical Release: Industrial Production, Board of Governers of the Federal Reserve System, July 18, 1985; and Resource Dynamics Corporation estimates and Battelle estimates.

glass, and primary metals. The incentive for electrification in these processes depends in part on temperature: the specific requirement and the availability of a technology to produce it. From a strictly theoretical perspective, the higher the temperature requirement, the higher the incentive for electrification, because electricity has no temperature limit.

Using electricity to raise a process temperature yields substantial benefits. First, virtually all of electric energy's heat content can service the process requirement. By contrast, if fuel supplies the thermal energy, a process can only use that fraction of the hot gases resulting from fuel combustion, and only if it is at a temperature above the process requirement. The rest of the thermal energy goes with the waste flue gas, which commonly carries away 25 to 75 percent of the fuel's heat content.

Second, reactions take place much more quickly at high temperature. The faster the process conversion occurs, the less heat is lost through conduction, convection, and radiation. Third, electric heat introduces no combustion products and associated impurities into the reaction, expends no capital, and wastes no energy. Finally, if a reaction's equilibrium favors high temperatures, the reaction will yield greater amounts of desired product.

Table 8-2 presents an examination of electricity use by the twenty industry sectors in 1980 and 1985. Over this five year period, total electricity use actually increased at an annual average of 2.7 percent for the metals fabrication and non-metals fabrication industries, while the metals production industries experienced a 4.3 percent per year drop in electricity use for the same period. Electricity market share ranged from 9.6 percent in SIC 32 (Stone, Clay, and Glass Products) to 40.0 in SIC 36 (Electrical Equipment).

Table 8-2
Industrial Energy Use and Electricity Use
By Industry Group (1980-1985)

SIC	INDUSTRY	1985 TOTAL ENERGY USE (TRILLION BTU)	PERCENT ELECTRICITY	TOTAL ELECTRICITY USE (BILLION kWh)		ANNUAL GROWTH PERCENT
				1980	1985	
PROCESS INDUSTRIES						
28	Chemicals & Allied Products	2,654	18.3	145.0	142.8	-0.30
26	Paper & Allied Products	1,366	21.1	76.0	84.7	2.19
20	Food & Kindred Products	1,029	15.9	43.3	48.0	2.08
29	Petroleum & Coal Products	1,037	13.9	37.7	42.3	-2.33
22	Textile Mill Products	286	30.8	26.1	25.9	0.15
21	Tobacco Products	21	23.8	1.4	1.6	2.71
	TOTAL PROCESS INDUSTRIES	6,393	18.4	329.5	345.3	-0.94
METALS PRODUCTION						
FERROUS PRODUCTION						
331	Blast Furnace, Basic Steel Prod.	1,011	18.0	82.8	53.2	-3.26
332	Iron & Steel Foundries	130	24.7	11.7	9.4	-4.28
339	Misc. Primary Metal Products	27	17.8	1.5	1.4	-1.4
	TOTAL FERROUS PROD.	1,168	18.7	96.0	64.0	-3.38
NON-FERROUS PRODUCTION						
333	Primary Non-ferrous Metals	343	61.2	84.9	61.6	-6.21
335	Non-ferrous Rolling & Drwg.	168	23.2	11.4	11.4	0.00
336	Non-ferrous Foundries	46	18.5	2.2	2.5	2.59
334	Secondary Non-ferrous Metals	35	9.7	1.0	1.0	0.00
	TOTAL NON-FERROUS PROD.	592	44.0	99.5	76.5	-5.15
	TOTAL METALS PRODUCTION	1,760	27.2	175.5	195.5	-4.35
MATERIALS FABRICATION						
35	Machinery Except Electric	321	32.1	30.7	30.3	-0.26
37	Transportation Equipment	388	30.4	30.0	34.5	2.83
36	Electric Equipment	280	40.0	27.2	32.8	3.82
34	Fabricated Metal Products	377	25.5	25.3	28.1	2.12
38	Instruments, Related Products	91	27.5	6.0	7.3	4.00
39	Misc. Manufacturing Ind.	43	27.9	3.6	3.5	-0.56
	TOTAL METALS FAB.	1,500	31.1	122.8	136.5	2.14

Table 8-2 (continued)

SIC	INDUSTRY	1985 TOTAL ENERGY USE (TRILLION BTU)	PERCENT ELECTRICITY	TOTAL ELECTRICITY USE (BILLION kWh)		ANNUAL GROWTH PERCENT
				1980	1985	
	NON-METALS FABRICATION					
32	Stone, Clay & Glass Products	1,183	9.6	30.8	33.4	1.63
30	Rubber & Misc. Plastics	304	33.9	21.8	30.2	6.74
24	Lumber and Wood Products	224	25.4	14.8	16.8	2.57
27	Printing and Publishing	112	37.5	9.7	12.4	5.03
23	Apparel, Textile Products	57	36.8	6.0	6.1	0.33
25	Furniture and Fixtures	52	28.8	4.0	4.5	2.38
31	Leather, Leather Products	14	28.6	1.4	1.0	-6.51
	TOTAL NON-METALS FAB.	1,946	18.2	88.5	104.4	3.36
	TOTAL MATERIALS FABRICATION	3,446	23.8	211.3	240.9	2.66
	TOTAL MANUFACTURING	11,601	21.4	716.9	726.7	-0.27

Note: Totals may not equal sum of components due to independent rounding. Source: U.S. Department of Commerce, Bureau of the Census, *1985 Annual Survey of Manufacturers, Fuels and Electric Energy Consumed*; Resource Dynamics Corporation and Battelle Estimates.

When electricity and energy consumption is analyzed using the four digit SIC classifications, a relatively small number of industries make up the overwhelming portion of both energy and electricity consumption. As shown in Table 8-3, twenty industries represent 53 percent of all fuels purchased by the industrial sector. As shown in Table 8-4, twenty industries represent 56 percent of all electricity consumed.

Table 8-3
Fuels Purchased by Selected Four-Digit SIC (1985)

SIC CODE	INDUSTRY	FUELS PURCHASED ($MILLIONS)	PERCENT OF ALL INDUSTRIES	CUMULATIVE PERCENTAGE	FUELS PURCHASED RANK	ELECTRICITY PURCHASED RANK
2911	Petroleum Refining	3258	11.6	11.6	1	4
3312	Blast Furnace & Steel Mills	2566	9.1	20.7	2	2
2869	Industrial Organic Chem. NEC	1878	6.7	27.4	3	7
2621	Paper Mills	1645	5.8	33.2	4	5
2631	Paperboard Mills	1067	3.8	37.0	5	11
3241	Hydraulic Cement	556	2.0	39.0	6	13
2873	Nitrogenous Fertilizers	551	2.0	41.0	7	23
2821	Plastics Materials & Resins	547	1.9	42.9	8	9
2819	Industrial Inorganic Chem. NEC	474	1.7	44.6	9	3
2865	Cyclic Crudes & Intermediate	409	1.5	46.1	10	22
3221	Glass Containers	372	1.3	47.4	11	26
3321	Gray Iron Foundries	347	1.2	48.6	12	17
3079	Miscellaneous Plastic Products	315	1.1	49.7	13	6
3711	Motor Vehicles & Car Bodies	303	1.1	50.8	14	14
2611	Pulp Mills	273	1.0	51.8	15	—
3714	Motor Vehicle Parts	271	1.0	52.8	16	10
2046	Wet Corn Milling	245	0.9	53.7	17	—
2824	Organic Fibers, noncellulosic	240	0.8	54.5	18	16
3353	Aluminum Sheet, Plate & Foil	220	0.8	55.3	19	24
3275	Gypsum Products	209	0.7	56.0	20	—
TOTAL MANUFACTURING		28,109	100.0	100.0	—	—

Source: U.S. Department of Commerce, Bureau of the Census, *1985 Annual Survey of Manufacturers, Fuels and Electric Energy Consumed;* Resource Dynamics Corporation and Battelle Estimates.

Table 8-4
Electricity Purchased by Selected Four-Digit SIC (1985)

SIC CODE	INDUSTRY	ELECTRICITY PURCHASED (MILLION KWH)	PERCENT OF ALL INDUSTRIES	CUMULATIVE PERCENTAGE	FUELS PURCHASED RANK	ELECTRICITY PURCHASED RANK
3334	Primary Aluminum	55,626	8.4	8.4	1	—
3312	Blast Furnace & Steel Mills	40,273	6.1	14.5	2	2
2819	Industrial Inorganic, NEC	34,128	5.2	19.7	3	9
2911	Petroleum Refining	33,211	5.0	24.7	4	1
2621	Paper Mills, Except Building Paper	28,520	4.3	29.0	5	4
3079	Misc. Plastic Products	22,958	3.5	32.5	6	13
2869	Industrial Organic Chem. NEC	21,054	3.2	35.7	7	3
2813	Industrial Gases	14,085	2.1	37.8	8	—
2821	Plastic Materials & Resins	11,845	1.8	39.6	9	8
3714	Motor Vehicle Parts	11,430	1.7	41.3	10	—
2631	Paperboard Mills	10,138	1.5	42.8	11	5
2812	Alkalies & Chlorine	9,676	1.5	44.3	12	23
3241	Hydraulic Cement	9,404	1.4	45.7	13	6
3711	Motor Vehicles & Car Bodies	7,742	1.2	46.9	14	14
2221	Weaving Mills, Manmade Fibers	6,738	1.0	47.9	15	—
2824	Organic Fibers, Noncellulosic	6,692	1.0	48.9	16	17
3321	Gray Iron Foundries	6,605	1.0	49.9	17	12
2421	Sawmills & Planing Mills, General	6,423	1.0	50.9	18	—
3662	Radio & TV Communication Equipment	5,962	0.9	51.8	19	—
3674	Semiconductors and Related Devices	5,550	0.8	52.6	20	—
	TOTAL MANUFACTURING	658,672	100.0	100.0	—	—

Source: U.S. Department of Commerce, Bureau of the Census, 1985 *Annual Survey of Manufacturers, Fuels and Electric Energy Consumed;* Resource Dynamics Corporation and Battelle Estimates.

Electricity's Future Role In The Industrial Energy Market

The industrial sector accounts for about 36 percent of both electricity and energy consumption in the U.S. economy. For individual electric utilities, industrial sales comprise not only a major part of system sales and revenues, but also represent an even greater percentage of earnings because of sales at a higher than system average load factor. Additionally, since industrial activity generates employment and indirectly sustains numerous residential and commercial sector activities within a utility service area, utilities are interested not just in analyzing industrial sales but also in ensuring that they continue to be a significant part of the service area's future.

The industrial market represents mixed opportunities for electricity. As shown in Figure 8-4, industrial energy consumption is decreasing as a share of total energy consumption for all sectors. Yet, electricity's share of the industrial energy market has increased from 14 percent in 1965 to 21 percent in 1985.

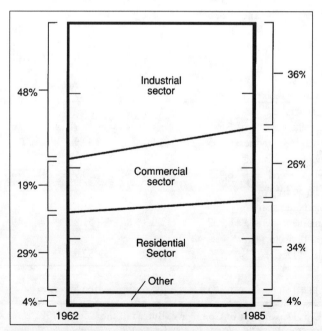

Figure 8-4. Distribution of electricity sales by customer class.
Source: U.S. Department of Energy, Energy Information Administration, *Monthly Energy Review*, May 1986

A number of studies have predicted that these trends will continue. As shown in Table 8-5, the 10– to 20–year forecasts made by the North American Electric Reliability Council, EPRI, and the U.S. Department of Energy's Energy Information Administration predict industrial growth rates at or below the growth rates of the economy as a whole. The projected growth rate for industrial electricity sales, however, exceeds the projected growth rates for both industrial production and the general economy.

The industrial energy market differs from the commercial and residential markets in a number of other ways. Analysts view the industrial sector as:

- Having a greater concentration of electricity consumption in a few, large customers.
- Being more vulnerable to overall business conditions and international factors.

Table 8-5
Projected Electricity Growth Rates

FORECAST	ANNUAL GROWTH RATES		RATIO OF GROWTH IN INDUSTRIAL ELECTRIC SALES TO		
	GNP	INDUSTRIAL ELEC. SALES	INDUSTRIAL PRODUCTION	GROWTH IN INDUST. PROD.	GROWTH IN GNP
NERC [1]	2.5	2.61	3.0	1.15	1.20
EPRI [2]	2.5	2.89	3.3	1.14	1.20
EIA [3]	2.7	2.44	2.74-2.95	1.12-1.21	1.00-1.10

Sources:
 (1) North American Electric Reliability Council, 1981 to 1991.
 (2) Electric Power Research Institute, 1981 to 2000.
 (3) Energy Information Administration, 1981 to 2000.

- Being more difficult to segment into "neat" homogeneous groups.
- Involving more decision makers in selecting energy investments and involving decision makers that may be located outside the plant and the utility's service area.
- Having at least a general familiarity with technology and engineering.
- Having greater opportunities for substitution among production inputs such as other energy sources, capital, labor, and materials.
- Placing a greater emphasis on economic hurdle rates to justify investments.
- Being more compatible with personalized or direct marketing approaches as opposed to mass marketing techniques.

Changes In Marketing To Industry

Today's electric utility industry faces a marketing environment that is significantly different from past years. In addition, the structure of the industry is changing dramatically. These developments affect how utilities market to all customers, particularly the industrial sector.

The industrial market represents major customer opportunities for large increments of load growth, conservation, and load management. The extent to which these opportunities are realized will be based on how well electric utilities control their costs, focus on markets with the greatest opportunities, and communicate the features and benefits of their services to end users. Adopting DSM strategies will comprise an important part of most future utility resource plans.(8-2)

A number of major changes in the industrial sector currently influence utility operations and sales. Some of these changes include:

- Changing economic base
- Increasing competition

- Developing new electrotechnologies
- Changing customer loyalties

Changing Economic Base

Three fundamental changes are taking place in the U.S. economy. First, industry now sells its products to global markets. Increasingly, consumer durable and nondurable goods must compete on a worldwide scale. This globalization has caused intense competition in some markets, leading a number of U.S. firms to relocate their manufacturing facilities to other regions or other countries to reduce production costs.

The second change is directly related to this shift toward global markets: the U.S. manufacturing sector must evolve to meet these new challenges or die. Primary metals, chemical and allied products, textiles, and a few other heavy industries have, like the electric utility industry, faced the challenges posed by increasing costs and aging equipment. For example, a number of electric utilities have experienced the effects of globalization in their own businesses. Local industries have closed plants or significantly reduced manufacturing employment; thus, their electricity demands have dropped either partially or completely.

The third major change centers on the transition of the U.S. economy from a product-based to service-based economy. Between 1984 and 1995, economists project that nine out of ten new jobs will come from service industries. Business services alone will represent 60 percent of the new jobs over the forecast period. Several manufacturing industries are expected to experience major declines in employment. Figure 8-5 summarizes past and projected employment levels for the time period 1960 to 1995.

Increasing Competition

The industrial sector can now choose from a growing number of competitors for energy services, including electric utilities, non-

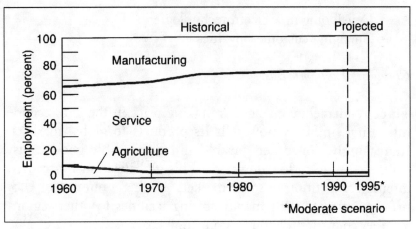

Figure 8-5. Employment trends, 1960–1995.
Source: U.S. Department of Labor, Bureau of Labor Statistics, US DL 85-478 November 7, 1985 extrapolated.

utility generators, and alternative fuels. In addition, energy conservation and management efforts are reducing overall energy use. As a result of these pressures and slower economic growth, the electric utility industry must compete with other energy suppliers to retain or increase its market share in the industrial sector. This competition has resulted from a number of institutional pressures, including the Public Utilities Regulatory Policies Act of 1978 (PURPA), tax policy, regulatory initiatives, and changing economic factors (e.g., the flattening of oil and gas prices). Regulatory policies in several industries (e.g., telecommunications, transportation, financial, and natural gas) continue to evolve.

Natural gas suppliers are under pressure from their industry to compete with each other by transporting surplus supplies via a third party to large end users located in another supplier's territory. Natural gas utilities have a large share of the commercial energy market and many plan aggressive marketing activities to protect that market share and offset recent gains by electricity. The gas industry strategies will focus on protecting the role of natural gas for process and space heating, implementing gas cooling, developing combined heating/cooling technologies, and promoting cogeneration. Industry efforts will be facilitated by decreased customer concerns about fuel availability and declining prices for natural gas and other fossil fuels.

Even moderate success of this campaign will significantly impact electricity sales to the industrial sector. Thus, the electric utility industry must begin examining DSM options that may reduce future capital outlays, make greater use of existing plants, and increase the value of electricity service to industrial customers.

Figure 8-6 plots the relative prices for electricity and fossil fuels. Prices, along with system efficiencies and productivity rates, combine to determine the relative cost-effectiveness of competing fuels.

Developing New Electrotechnologies

Electricity represents a basic ingredient in most new productivity improving technologies and offers a number of inherent technical, economic, social, and policy advantages. In particular, elec-

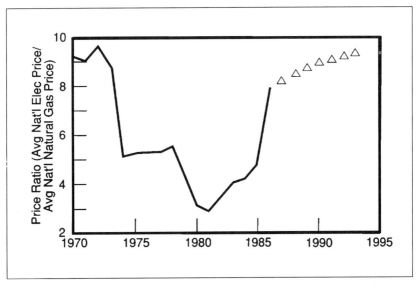

Figure 8-6. Price competitiveness of electricity.
Source: A. Faruqui, P. Gupta, and J. Wharton, "Ten Propositions in Modeling Industrial Electricity Demand," in Adela Bolet (ed.) *Forecasting U.S. Electricity Demand*, (Boulder, CO: Westview Press), 1985.

tricity presents unique advantages for the industrial user, including:

- Enhanced technical features—efficiency, controllability, intensity, and flexibility
- Reduced environmental concerns, capital requirements, and energy consumption
- Increased rates of production and raw materials
- Improved product quality and yield

As an indication of electrotechnology trends, Figure 8-7 provides rough estimates of the impacts that selected electrotechnologies will have on industrial electricity sales during the year 2000.

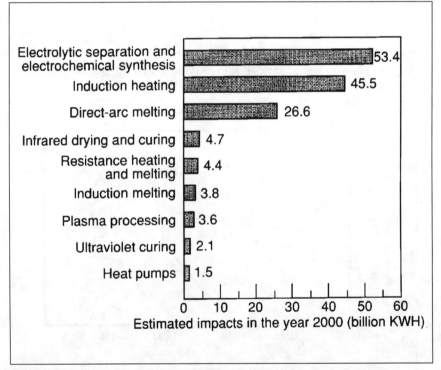

Figure 8-7. Industrial electrotechnologies with significant energy impacts.
Source: *Industrial Electrotechnologies and Electrification – Volume 1: Electrotechnology Reference Guide*, EM-4527, developed by Resource Dynamics, Inc. for EPRI, April 1986.

Changing Customer Loyalties

Most industries, including the electric utility industry, feel that customers are not as loyal as in the past. Long-term sales stability results from having a group of loyal customers. Customer loyalties are influenced by a number of factors including:

- Prior experience
- Perception of utility concern for customers
- Prices
- Availability of substitutes
- Bundling/unbundling of services
- Product attributes
- Process requirements

The electric utility industry will no doubt have to work hard to maintain and gain additional customer support. Utilities must devise balanced marketing activities that promote attractive corporate images, value-of-service, caring attitudes, and DSM programs.

Summary

The industrial market is characterized by a diversity of processes, end uses and business types, making it the least homogeneous of all sectors. Industrial customers are also highly vulnerable to changes in domestic and international business conditions and have greater opportunities for fuel substitutions. At the same time, these customers consume about 36 percent of the electricity and energy used in this country. The industrial market offers the electric utility industry greater competitive opportunities (and risks) as well as the potential for technological advancement.

References:

8-1 (a) "Propositions in Modeling Industrial Electricity Demand," *Forecasting Electricity Demand,* A. Faruqui, P. Gupta, and J. Wharton.

(b) *Industrial Electrotechnologies and Electrification—Volume 1: Electrotechnology Reference Guide*, EM-4527, developed by Resource Dynamics, Inc. for EPRI, April 1986.

(c) *Descriptive Classification of Electro-Technologies and Manufacturing Productivity Implications* (forthcoming), RP-2381-02, developed by Mathtech, Inc. for EPRI.

(d) *Proceedings: Forecasting the Impact of Industrial Structural Change on U.S. Electricity Demand*, EA-3816, published by EPRI, December 1984.

8-2 *Demand-Side Management—Volume 5: Industrial Markets and Programs*, EA/EM-3597, prepared by Battelle Columbus Division and Resource Dynamics Corporation for EPRI, March 1988.

CHAPTER 9
Case Studies in Four Industries (9-1)

This chapter briefly examines the historical impacts of electrotechnologies on four different industries: pulp and papermaking, glassmaking, petroleum refining, and agriculture. It discusses traditional manufacturing methods in each industry and then explains how new electric powered technologies save money and improve both energy and production efficiencies.

Electrification in Pulp and Papermaking (9-2)

The pulp and papermaking industry employs both mechanical and chemical pulping methods. The traditional method of mechanical pulping uses equipment that debarks roundwood, presses it against a rotating grindstone in a bath of water, and tears apart the fibers. This action produces an aqueous slurry. The brute-force wood shredding of this process produces short fibers and ultimately yields paper with poor physical properties. To improve paper quality, pulpmakers blend this pulp with different amounts of higher quality chemical pulp to make newsprint, catalogs, container board, and tissue. The industry makes about 11 percent of U.S. pulp this way.

173

Despite the low quality of conventional roundwood pulp, the paper industry uses it whenever possible because it is inexpensive. Its low cost results from its efficient production: about 95 percent of the roundwood fed to the grinder ends up as pulp. In addition, the manufacture of this pulp requires no chemicals, thus eliminating their initial costs as well as the costs associated with reprocessing spent liquor and treating noxious wastes. The advantages of mechanical pulping—high yield and low environmental impact—provide strong incentives to improve the quality of mechanically-produced pulps so the industry can use them more.

In the traditional method of chemical pulping, equipment first reduces roundwood into wood chips. Then, the chips undergo a high pressure/temperature treatment in a sulfur-based chemical bath. This bath softens the chips, modifies the lignin, and readily disintegrates the chips into fibers. Equipment then washes the resulting pulp to recover lignin and pulping chemicals. Depending on the ultimate use, processors may bleach this chemical pulp to improve physical properties, color stability, and light reflectivity.

Chemical pulp offers good paper strength and color stability because the natural wood fibers are separated without being physically damaged (i.e., shortened) by mechanical grinding. Also, the process removes much of the lignin. Papermakers use chemical pulp for all grades of paper and paperboard products, but it is costly. The process uses only about 50 percent of the wood resource and requires additional costs to recover pulping chemicals, the energy contained in lignin wastes, and treat the effluents.

The pulp and paper industry currently seeks some middle ground between these processes. The nonchemical pulping method offers low cost and low to moderate quality products and is environmentally acceptable. The chemical method exhibits the opposite characteristics. Industry researchers are developing alternatives to both processes with success, and ultimately a much improved wood-pulping process will probably replace the traditional processes now in common use. Electricity is playing a major role in these new advancements, and thus the industry's electricity purchases continue to grow—from 680 kWh per ton in 1972 to 790 kWh per ton in 1981.

Improvements In Mechanical Pulping

In pulping operations, for example, the industry increasingly uses chemimechanical and thermomechanical pulping techniques and combinations of the two. One alternative mechanical pulping process centers on improving grinding technology. Producers can pretreat the wood thermally, chemically, or both before grinding it. Electricity powers virtually all of the debarking, chipping, and grinding equipment used in mechanical pulping. For example, typical refiners used in newsprint plants use 2,200 HP motors. Even the thermal energy requirement for thermomechanical pulping is, in the future, likely to be provided by grinders designed to deliver pressurized steam raised from the heat generated in the grinding process.

Thermo- and chemimechanical techniques call for chipping the roundwood first and then presteaming it before grinding. Since this process softens the wood, the fibers separate more easily during grinding and sustain less physical damage. The resulting fibers are often twice as long as those made by conventional grinding. The yield from thermo- and chemimechanical processes is about 97 percent, compared with a yield of only 50 percent from conventional kraft processing. In addition, these processes generate little or no black liquor, thereby avoiding the costly cleanup of the sulfite and sulfide waste streams that result from kraft pulping. Capital costs are about 40 percent of equivalent kraft process plants.

These thermo- and chemimechanical pulps have better physical properties and thus broader papermaking applications. Paper producers can make newsprint entirely from thermomechanical pulp instead of the traditional blend of about 20 percent chemical pulp and 80 percent groundwood. Many paper products use some percentage of mechanical pulp, so the higher quality thermomechanical pulp offers even broader applications and new markets. For example, this pulp is now used in Europe as filler in disposable diapers.

The doubling of fiber length that results from thermomechanical pulping means that, with some chemical pretreatment, the industry can produce usable pulps from low density hardwoods such as aspen and poplar. Since hardwoods contain less lignin than do softwoods, thermomechanical hardwood pulp offers a higher quality than conventional groundwood pulp made from softwoods. In addition, hardwood pulps absorb liquids better than many softwood pulps.

These improvements in mechanical pulping techniques extend the resource base to some hardwoods as well as expand the market for high yield pulps for a wide variety of paper products. This pulping process offers a relatively environmentally benign method of production as well as results in lower capital, raw material, and operating costs of production than for the chemical pulp that is displaced. For these reasons, thermomechanical pulping is the fastest growing sector of the pulp and paper industry.

Advances In Chemical Pulping

Modern thermo- and chemimechanical processes provide higher quality pulps than conventional methods, however these processes still have limitations. The wood fiber sustains some damage in grinding and the processes cannot remove all of the lignin. Chemical pulping still produces a higher quality pulp (longer wood fibers and better physical properties and color stability because more lignin is removed). However, the capital and operating costs incurred in purchasing and recovering chemicals raise the cost (and value) of the pulp and the effluents from reprocessing impact the environment.

For example, to produce a bright, color stable paper, processors must bleach the chemical pulp from the digester. The bleaching process includes a series of chemical treatments using chlorine. The process produces effluents that contain corrosive chlorides and toxic chlorinated organic compounds unsuitable for recycling or release to natural bodies of water. Conventional pulping practice treats the effluent biologically in aerated lagoons. This treatment does reduce the level of chlorinated organic compounds but does not improve the dark color of the effluent.

All research to improve chemical pulping processes seeks to solve environmental problems. In addition, all approaches involve the use of electricity. For example, almost all approaches substitute oxygen for part or all of the sulfur and chlorine. Any approach that displaces sulfur implies greater use of electricity because electricity must produce the various forms of oxygen.

The pulping industry has adopted one "evolutionary" approach to reducing the levels of toxic chemicals in the process. This approach incrementally reduces the use of chlorine for bleaching, thus allowing the industry to progressively accommodate more restrictive environmental regulations. The industry ultimately wants to produce a bleach plant effluent suitable for reuse in pulping. Completely replacing chlorine with oxygen would meet the goal, but oxygen cannot completely bleach the pulp to the same degree as can chlorine.

A revolutionary approach studied by the industry seeks to replace the "conventional" kraft process and eliminate both sulfur and chlorine. This process chemically pulps wood chips using oxygen, or delignifies thermomechanical pulps by treating them with oxygen and alkaline solutions.

In summary, as these examples show, the distinctions between chemical and mechanical pulping are becoming blurred. Wood pulping becomes increasingly electrified since the new pulping processes depend on mechanical fiberization and on the use of electrically-produced chemicals such as oxygen, chlorine dioxide, and ozone. The industry will continue to adopt these evolving processes, and the resulting product will be characterized by higher average fiber yield and lower environmental impact.

Paper Drying

In terms of primary energy used (i.e., including waste heat from electricity production), electricity provides one-third of the energy required to make paper products from pulp. Thermal energy, either in the form of steam or hot air, uses the remaining two-thirds of the energy to evaporate water. While the drying process removes 98 percent of the water mechanically, the removal—using heat—of last the 3 pounds of water per pound of paper product uses most of the energy. This inefficiency, along with the relatively large amount of energy the process requires, provides the industry with a strong incentive to improve the energy efficiency of the evaporative drying portion of the papermaking process.

One technology improvement, called radio frequency (RF) heating (sometimes called dielectric heating) adds heat to the evaporation process without increasing the heat losses. RF heating adds an increment of energy to the wettest part of the paper surfaces. This technology results in stronger, more uniformly dried paper, reduces the need for additional drying, and saves energy. Research shows that RF drying offers a "bonus effect": it removes more moisture than can be accounted for by the total added electric energy. This bonus may occur because RF heat moves moisture from the paper interior to the paper surface where the conventional steam rolls evaporate it more effectively. In some instances, the moisture is vaporized so rapidly that it forcibly ejects water particles from the paper surface without evaporation. Thus, RF drying permits a 20 percent increase in machine throughput. This increase is larger than expected from the amount of energy input, and therefore, the industry can justify a proportionately higher cost for energy in that form, especially if the process increases productivity and improves paper quality. Today, applications of RF paper drying provide about 30 percent of the evaporative heating requirement and could meet all needs.

Impulse drying, although not yet commercialized, also offers a promising improvement in energy efficiency. In impulse drying, induction or electric IR heaters heat press rolls to about 700°F. Paper comes in contact with these hot metal rollers under pressures of 400–700 psi for 15–100 milliseconds. This application of heat and pressure can potentially reduce energy requirements for the initial drying step by as much as 75 percent as well as the paper's tensile strength by as much as 35 percent.

Ancillary Electric Technology In The Pulp and Paper Industries

The preceding sections discussed the expansion of electrotechnologies into the basic business of producing wood pulp and paper. Additional electrotechnologies meet other needs in the industry: membrane separation processes concentrate spent sulfite liquor and purify boiler feedwater; centrifugation separates solids from wastewater streams; ultrasonic systems clean paperdry felts; and computerized process systems control a variety of pulp and papermaking operations.

Conclusion: Implications of the Changes on the Industry's Electricity Demand

The complexity of the pulp and paper industry means there is no "template" for electrifying the industry. Different paper products require different mixes of wood pulp which, in turn, comes from different processes (e.g., some require bleaching and some do not). Fiber sources used for pulping differ in their composition and characteristics. Accordingly, a simple model of energy and materials only helps industry planners understand general relationships and implications of process change, not plan specific strategies.

Table 9-1 provides information about the general energy and raw material requirements of making pulp and paper. The information covers basic production of pulp, paper, and paperboard but not the industries that convert these materials into, for example, boxes or envelopes. Producing each ton of paper requires 1000 kWh of mechanical energy plus 22.5 MBtu of heat energy. Each ton of new pulp comes from nearly two tons of wood fiber.

Estimates made for the years from 1975 to 1986 show that the industry's total energy use per unit of output declined approximately 9 percent. At the same time, the electricity share increased 22 percent. In 1986, the pulp, paper, and paperboard industry purchased some 43 billion kWh of electricity and over 800 trillion Btu of various fuels in other forms.(9-3)

Table 9-1
Typical Energy Requirements per Ton of Paper
Weighted by Process Fraction

Pulping Process	Fraction of Production	Process Contribution to Energy Required per Ton	
		kWh Electrical	MBtu Thermal
Wastepaper recycle	20	140	2.6
Chemical	70	665	18.9
Mechanical	10	195	1.0
		1000	22.5

Several process changes will substantially increase electricity consumption in the years to come. These changes include:

- Shifting from chemical to mechanical pulping (increases electricity use)
- Using more recycled material (decreases electricity use)
- Adopting RF drying technologies (increases use)

- Replacing chemicals with oxygen for chemical pulping (increases use)

If all these changes proceeded at a 1 percent annual rate, the industry could increase electricity use per ton of output about 2 percent per year. However, the industry would need to work quickly to adopt commercial oxygen pulping and delignification in the U.S. Also, overall process energy requirements will decrease as the existing and new processes are improved. For example, with oxygen pulping, less mechanical energy is required to refine (i.e., beat) the pulp.

Glassmaking (9-4)

Glassmaking using gas-powered melting systems has been practiced worldwide for over a century. Electric-powered glass melting, on the other hand, is a relatively new, yet more efficient, technology.

Electric Melting

Glassmakers in Sweden first attempted to melt glass using electricity in 1925. Early efforts relied on passing an electric current through molten glass; while innovative, this method failed to produce a consistent, good quality product. By the early 1950s, glassmakers had corrected these technical problems as electricity-based methods evolved. Each generation of electric powered glassmaking equipment improves in efficiency. A large (typically 120 tons per day) all-electric furnace uses electricity at an efficiency rate of 66 percent. For small all-electric furnaces (about four tons per day), the efficiency rate drops to about 34 percent. By the mid-1980s, almost 250 trillion Btu of energy was used by the U.S. glass industry—nearly 80 percent of it natural gas, about 18.5 percent electricity, and the remainder, oil.

The melting operation portion of the glassmaking process consumes 80 percent of the energy required for glass production. Thus, electric processes hold great energy saving potential. Electric efficiency improves when glassmakers use a process that electrically "boosts" fuel-fired gas furnaces. The very high efficiencies of this electric boosting process result because the additional electric energy melts extra glass. This technology also reduces the amount of energy used per ton of glass melted, even when considering all the fuel used to generate and deliver electricity. Some factories have increased their output 60 percent by adopting the electric boosting process alone. Although factories may reduce overall fuel use with this process, the energy cost per ton is often higher because of electricity's higher cost.

Data that compares energy use between gas and all-electric melting suggests that 1 Btu of thermal energy delivered as electricity is equivalent in melting capability to 2.5 to 4 Btu of energy delivered as gas. Since about 3 Btu of energy as fuel is needed to generate a Btu of thermal energy as electricity, all-electric melting would compete with gas if electricity cost no more than three times the cost of thermal energy as fuel. However, thermal energy from electricity has historically cost much more than three times the cost of thermal energy from gas. However, the option of electric boosting substantially changes the economics, offering overall process energy efficiency with costs somewhere between all-fuel and all-electric melting.

For example, a 100 ton-per-day glass melting facility using gas can increase production to 160 tons per day through electric boosting. In addition, electric boosting raises the overall thermal efficiency of this plant from 25 to 35 percent. This higher efficiency rating is much closer to the efficiency of all-electric melting, but electricity accounts for only 13 percent of the total thermal energy supplied. The increased production through electric boosting also effectively reduces the capital cost per unit of output. This cost reduction explains why the industry is adopting electric boosting and mixed melting, particularly as the cost of gas increases relative to the cost of electricity. It also explains why all-electric melting has only been adopted for more or less special situations where electricity is unusually low in price (e.g., regions rich in hydropower) or where the special advantages of electric melting justify its high energy cost.

All-electric melting processes effectively solve the industry's difficult emissions problem. An all-electric melting process releases only 4 percent of the emissions released by fuel melting, and these emissions are almost entirely comprised of water vapor and CO_2. In addition, if the electricity used comes from sources that present little or no emissions (e.g., nuclear and hydro power) or from sources in which the emissions are more readily controlled (e.g., coal combustion), the process reduces glassmaking's overall environmental impacts.

The use of electricity for glass melting has greatly expanded over the past thirty years. About half of the container glass manufacturers in the U.S. now use electric boosters, and at least 100 all-electric furnaces, ranging in size from 4 to 140 tons per day, operate worldwide. However, the relatively high cost of electricity limits the use of all-electric melting to areas of the country with lower electricity prices.

Gas Melting

Gas powered glass melting furnaces typically have fuel energy efficiencies ranging from 15 to 25 percent. Energy efficiency depends strongly on two factors: the operation's size and the production rate. The scale of operations determines the furnace surface-to-volume ratio (i.e., it is higher for small tanks than for larger ones; thus, the fraction of total energy lost through furnace walls and flue gas systems is higher for small tanks than large tanks).

The production rate for a given melting tank also determines efficiency, since a system loses energy through the furnace walls and flue gas system at a constant rate over time. In addition, the rate is proportionately higher for a low production rate than for a higher one. For example, the amount of energy used to keep the tank hot and ready for production is commonly about half the amount required during production. (These functional relationships between energy efficiency and scale of operations and production rate apply to electric melting also, but not to the same degree because energy losses are much smaller for all-electric melting.)

Electric Heat For Glass Conditioning and Annealing

Gas is traditionally used for two other process heat requirements in glassmaking:

- The "forehearth," or conditioning trough, which conveys the molten glass from the melter to the glass-forming machines.
- The annealing furnace where the shaped glass products can, if necessary, be reheated and cooled to room temperature under controlled conditions to reduce or eliminate residual stresses in the final product.

These operations typically require 5 to 10 percent of the gas needed for melting. Increasingly, glassmakers are replacing gas systems with electric ones and achieving good results. Since these operations require careful temperature control—such as that offered by electric heat systems—glassmakers have adopted them more readily.

The electric forehearth uses 10 percent as much energy as does its gas-fired counterpart and offers a 2 to 5 percent increase in electric operations production efficiency. This small improvement can save nearly as much energy at the melting stage as does electrification of the forehearth operations. Thus, the total energy savings through electrification of the forehearth can be as much as 140 percent of the energy used for gas operations of the forehearth. Of course, an improvement in production efficiency translates into reduced product cost in all other operations as well. Other benefits of electrifying the forehearth include faster response to varying production demands and an improved working environment (because the electric forehearth structure is cooler, there are no waste gas fumes, and the noise level is lower).

Similarly, electric annealing furnaces offer energy savings of up to 90 percent. These furnaces eliminate the flue gas and equipment energy losses, since their surfaces are fully sealed (they do not need air and gas supply or discharge lines). In addition, if a factory needs space heating, it can duct the waste heat (since it is free of flue gases) and use it to achieve additional savings. Other claimed benefits include a clean working environment, less maintenance, accurate temperature control, high product quality, and a low reject rate.

Additional Benefits of Using Electricity in Glassmaking

Electricity is replacing other fuels in the glassmaking process for three reasons: falling electricity costs, lower environmental impacts, and the flexibility offered by electric powered processes. First, the basic cost of energy is driving many manufacturers to switch to electricity. Relative to natural gas prices, electricity prices are dropping.

Second, the growing need to reduce pollution favors electric processes. All-electric glass melting eliminates the loss of certain ingredients from the molten glass mixture. Glassmakers add ingredients such as oxides of boron, fluorine, lead, and phosphorous to batches of molten glass to impart particular properties, to improve workability of the glass, to remove bubbles, or to reduce the melting temperature. In fuel-fired systems, up to 50 percent of these materials escape from the molten surface. Once released, they may clog airflow passages (and reduce efficiency) or escape and add to atmospheric pollution. In either case, costs increase because the materials are lost. If captured in a suitable form (e.g., by electrostatic precipitation), systems can recycle most of the material. However, factories incur operational costs to recover and recycle the material. In addition, the composition of the product glass may not be consistent from batch to batch.

All-electric melting essentially eliminates this loss of materials. In this process, the molten glass is covered with several inches of unmelted batch, which serves as a cool filter and captures the volatile materials trying to rise from the molten surface. Material costs are reduced approximately 3 percent and since batch materials do not escape, the operational problems associated with managing batch materials do not arise.

Third, electric glass melting also provides recognized operational benefits that are difficult to quantify because they are intermittent or operational in character. Electric melting permits faster heat up and cool down. Thus, if quality problems arise because of errors in batch composition or in melting and refining, the problem is short in duration. Similarly there is less product lost in restarting operations after holiday cool down. Also with electric heat, colored glasses can be produced in deep tanks not specifically designed for their production, thus improving flexibility in operations.

Finally, the benefits of an all-electric glassmaking operation are yet largely unrecognized because most manufacturing experience has been with electric boosting or mixed melting. All-electric melting reduces an operation's noise level, offers dust-free operation, and provides easy access to the melting tank for delivering raw materials and for visual monitoring of the melting operation.

Petroleum Refining [9-5]

While electricity remains a small portion of the total energy used in petroleum refining, the amount used to refine a barrel of oil doubled between 1954 and 1981 (Figure 9-1). The entire increase comes from purchased electricity (Figure 9-2); refiners' self-generated electricity remained nearly constant over this period (also shown in Figure 9-2). During roughly the same time, per-barrel fuel use decreased 24 percent between 1962 and 1982 (Figure 9-3), with more than half of the decrease occurring between 1974 and 1977.

Analysts interpret these trends in several ways. The data could suggest refineries began directly substituting electricity for fuels. Or, it could indicate a combination of two separate forces—one

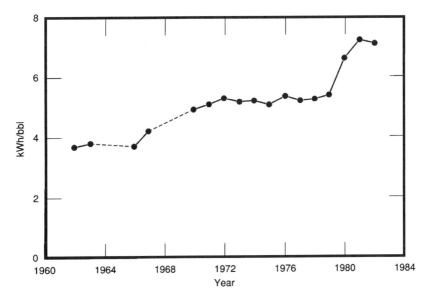

Figure 9-1. Growth of electricity use in petroleum refining.

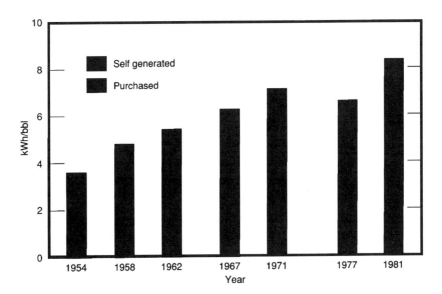

Figure 9-2. Purchased versus self-generated electricity used in petroleum refining.

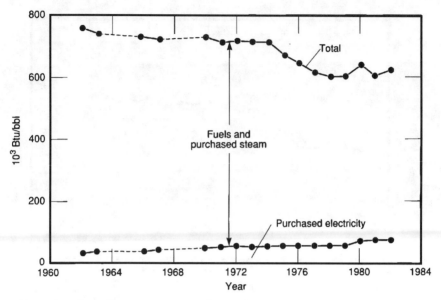

Figure 9-3. Overall energy use in petroleum refining.
The fuels data in Figure 9-3 includes the "nonpurchased fuels": still gas, petroleum coke, and other components of refiners' raw materials that are consumed as fuels. Use of nonpurchased fuels remained near 350,000 Btu per barrel throughout the period under study. As was the case with electricity, most of the change in fuels consumption occurred in the purchased component. In 1962, 1 Btu out of every 9 purchased was in the form of electricity; by 1982, the ratio was 1 out of 3.

fuel saving and the other electricity using. These hypotheses fail for three reasons: 1) decreases in fuel use and increases in electricity use are not occurring simultaneously, 2) the apparent amount of energy displaced by electricity seems implausibly high for direct substitution (approximately 50,000 Btu per kWh), and 3) there is no evidence of major new processes that would account for indirect substitution of this magnitude. Yet the data shows that some fuel substitution is occurring. Both interpretations offer some valid possibilities, but neither by itself is sufficient to explain the energy use trends in refining.

Product Quality

Two major forces affect total energy use and the fuel–electricity mix in petroleum refining: prices and product quality. Price considerations induce refiners to use less energy and to substitute electricity for fuels; quality improvements require refiners to expend more energy, especially electrical/mechanical energy.

Product quality has changed significantly over the past several decades. For example, technological advances in the performance of the internal combustion engine required motor gasoline quality to improve substantially during the 1950s and early 1960s. Octane ratings and volatility increased while sulfur content was halved. Beginning in the late 1960s, environmental regulations required refiners to reduce the lead alkyd used for octane boosting. Consequently, the lead content of leaded gasoline was halved, and unleaded gasoline made up half of all production by 1981.

Another example of changes in product quality is in sulfur recovery, which increased dramatically after 1970. Environmental requirements to reduce sulfur emissions forced refiners both to clean their own emissions and to put cleaner fuels on the market. At the same time, the sulfur content of crudes was increasing. Other environmental restrictions on solid, liquid, and gaseous effluents forced refiners to clean up emissions. Rising contaminant levels in crude oils, such as heavy metals, forced additional decontamination.

The most complex of the quality changes in refining output centers on the product mix. It constantly shifts in response to changes in demand for specific products and the incentives (regulatory and other) to import products rather than crude oil. The refinery product that has experienced the greatest change over the last three decades is residual fuel oil—the leftovers of petroleum refining. Residual production decreased steadily until the early 1970s, then rose quickly to peak late in the decade before heading downward again (Figure 9-4).

The direction of these changes in refining have virtually all been toward higher quality. Refiners produced better and cleaner outputs from poorer and dirtier inputs (the temporary shift toward

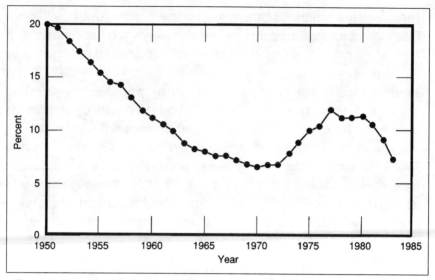

Figure 9-4. Residual fuel oil production as percent of crude oil.
Resid production can be reduced by using lighter crude oils or by additional processing, which upgrades residual oils to higher valued products. Added processing accounted for most of the pre-1970 decline. During the 1970s, many simple refineries were built, so the average level of processing declined. Refiners also were forced to resort to heavier crudes, further raising resid production. After 1980, additional processing of resids was reemphasized.

residential production during the 1970s was the only anomaly). Additional processing yielded higher quality processing: more cracking and coking, reforming and alkylation, hydrotreating and waste treating.

This additional processing uses more energy. Because higher quality products require more sophisticated chemical processes, which in turn involve a greater need to compress gases, a higher proportion of refinery energy must be in mechanical form. Thus, the compression of gases, which uses motors and pumps, can become a significant energy consuming refining activity.

By American Petroleum Institute (API) reckoning, if the oil refining process had been the same in 1972 as it was in 1982, then 1972 consumption would have been *17 percent higher* than reported. The increased processing requirements are allocated in Table 9-2 below:

Table 9-2
**Processing Requirements
(in 103 Btu per barrel):**

Lead phaseout and higher octane	18
Increased desulfurization	11
Product mix changes	28
Other processing adjustments	3
Major capacity additions	11
Processing of effluents	12
Throughput effect	10
Miscellaneous	17
Total	**110**

Thus, API estimates the amount of energy conservation achieved for the period 1972–1982 was the 1972 consumption rate (652) plus increased process requirements (110) less the 1982 consumption rate (587), or 175,000 Btu per barrel.

The increasing quality of products can help explain the bias toward electrification in refining over the past several decades, since product quality improvements require more energy. Clearly, the same barrel of crude oil was processed more substantially in 1982 than in 1962. The apparent reduction of approximately 20 percent in energy use over the two decades understates the true energy savings that occurred, because a barrel of oil in 1982 was processed much more extensively. Using API estimates for the 1972–82 decade, apparent savings were 10 percent, but real savings were 23 percent.

Interestingly, between the 1950s and 1980s, electricity prices have been falling, relative to fuels—this fact is not necessarily well known. In 1954, a Btu of electricity delivered to refiners cost 14.5 times as much as a Btu of energy as fuel. In 1971, the ratio had fallen to 7.9 and, by 1981, it dropped to 3.6. Low gas and oil prices in the mid-1980s slightly increased the ratio; however, analysts expect it to resume its drop in the next decade. This drop over the long term has supported the growing use of electricity in petroleum refining.(9-3)

Electricity-Using Refinery Technologies

Refiners use several technologies by which they directly substitute electrical or mechanical energy for thermal fuel energy.

Mechanical drive

A number of factors favor the use of electric motors over steam turbines: electricity's falling cost (relative to fuels' costs); the recovery and reuse of low-temperature steam and heat (which reduce the need for steam exhausted from back-pressure turbines); and advances in electronically controlled technologies. This last factor will play an increasing role well into the twenty-first century.

So far, little evidence of widespread conversion to motors appears in existing units. Because of the current sunk cost advantage of existing steam turbines, most refiners will not replace them until major refurbishing is required. In new units and retrofits, however, motor drives are now commonplace in applications formerly dominated by steam turbines.

Vapor recompression heat pumping

In some processes, equipment can compress exhaust vapors to recover and reuse their heat. Vapor recompression is, in effect, an open cycle heat pump. Mechanical compression raises the temperature of exhaust vapors so that the heat of vaporization, normally lost in a condenser, can be reused. The process directly substitutes electrical/mechanical energy for fuel/thermal energy. While heat pumping makes economic sense only for applications employing small temperature differences, refiners are finding a growing number of such applications.

Gas separation

New mechanically driven gas separation technologies are finding widespread use in refining and related industries. As an example, hydrogen recovery reduces the cost of many separation tasks and thus makes it feasible to economically produce many previously economically marginal products. Hydrogen recovery from refinery waste gas streams is already a major application, providing more hydrogen for product upgrading without requiring additional natural gas reforming.

Electronic controls

Refiners that adopt modern electronic control systems and minicomputers increasingly substitute electricity for fuels. A negligible amount of electricity can carry and manipulate information about process variables, thus allowing greater process control and resulting in remarkable energy savings. For example, an electrical spark flare igniter eliminates the need to fuel a pilot light on each flare stack, saving much fuel by using a tiny quantity of electricity.

Agriculture (9-6)

Agricultural practice and the technologies supporting it has moved through three phases. It began as a system totally dependent on human, animal, and solar energy, then shifted in the middle of this century to one heavily reliant on motor fuels. Finally, as we move into the twenty-first century, agriculture uses increasing amounts of natural gas and electricity.

The growing importance of electricity in recent years derives in large part from its use for delivering irrigation water, producing agricultural chemicals, and raising livestock. Irrigation, even in naturally watered areas, is responsible for large productivity gains resulting from full and efficient use of fertilizers, chemicals, and high-yield seed stock.

Electricity came late to the farm. By 1930, central station electricity powered only 9.5 percent of the nation's farms. The few farms that relied on electricity fell into special categories: farms either on the edge of urban communities (and thus close to the existing electrical network) or those specializing in dairy or poultry operations. These latter farms required substantial amounts of motor power for feed and water handling, milking machines, and milk and egg refrigeration.

High costs provided the major obstacle to further electrifying rural America. These costs included extending electric lines to serve isolated farms which used very small amounts of electricity. Rural areas had about five farms per road mile in 1930; estimates showed costs would have included an estimated line cost of $2,000 per mile in addition to about $100 per house for wiring and the cost of appliances to use the electricity. Even at fairly high electric rates of 4 to 12 cents per kWh, annual use had to be 600 to 1,000 kWh to justify the capital cost of electrification; the Alabama Power Company's test project showed that electricity use on the average cotton farm was only 240 kWh per year.

Despite its poor economic justification, early leaders of the move toward electrification recognized that its benefits would "result in a higher standard of living which cannot be figured in dollars and cents or in kilowatts." Elsewhere in the world, perhaps because of more populous rural areas, industrialized nations moved quicker to bring electricity to farms and had long since justified heavy farm subsidies to accomplish it. The Weimar Republic claimed 60 percent rural electrification in 1927. France had achieved 71 percent by 1930. Other countries far ahead of the U.S. included Finland with 40 percent, Sweden and Denmark with 50 percent, and Czechoslovakia with 70 percent in the early 1930s. Inevitably, political pressure would build to provide the rural dweller "an even chance with his city brother in the comforts of life." In July 1936, Franklin Roosevelt created the Rural Electrification Administration (REA).

The REA chartered farm cooperatives to purchase bulk power and help them finance the construction of distribution systems and wiring of rural homes. Government hydropower projects like the Tennessee Valley Authority (TVA) provided power at low cost. The Reconstruction Finance Corporation made $100,000,000 available for farm electrification during its first two years of operation, and Congress was authorized to appropriate funds up to $40,000,000 per year for the next eight years. Families needing money for wiring and appliances could apply for small individual loans. In 1936, the government interest rate was 3 percent, a rate lower than the yield on utility securities, which for 1935–36 varied from 3.25 to 3.88 percent.

The REA devised the "Arkansas plan" for impoverished Southern farmers. The plan included membership ($5), house wiring ($10), an iron ($3), and a radio ($7). Farmers paid 89 cents per month for the plan, on installment, as well as a $1 minimum monthly power purchase for the first 11 kWh. Their monthly bill could total as little as $1.89. Penniless farmers worked as laborers on construction crews to provide payment and permit those at the bottom of the agricultural ladder some benefits of modern living. Farming families used electricity first for lights, an iron, and a radio. Indoor plumbing, which required a pressurized source of water, was more expensive and came later. For expensive appliances, Southerners preferred refrigerators, whereas Northerners opted for washing machines.

The electrification program moved slowly because of higher federal priorities imposed by World War II and related shortages of manpower, fuel, and copper. By 1944, 55 percent of U.S. farms still had no service, but the program was popular, and Congress expanded appropriations after the war. The highest annual appropriation reached ten times higher than pre-war funding rates. By 1953, 2,544,000 farms, about half of all U.S. farms, were connected to REA-funded systems and for all practical purposes, American agriculture was electrified (Figure 9-5).

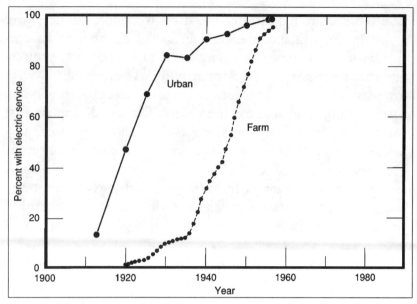

Figure 9-5. U.S. farm electrification.

Electrification directly affected patterns of agriculture, particularly in the South. For example, the U.S. Department of Agriculture reported the conversion of many cotton farms into dairies. Similarly, the South, by 1955, produced 69 percent of broiler chickens compared with 27 percent in 1935.

However, household operations consumed the bulk of the electricity and agricultural operations benefited through the resulting improvement in living standards. TVA reported that electric lights added the equivalent of ninety-one eight-hour days per year. Electric lighting is also credited with reducing home accidents, protecting eyesight, promoting cleanliness, and providing psychological benefits. Indoor plumbing and the electric washing machine saved an estimated forty-five eight-hour workdays per year and reduced farm sickness caused by polluted water. Refrigerators improved the farm diet and reduced the high incidence of food spoilage and staphylococcus food poisoning. The radio brought information on weather and markets. But more importantly, according to Brown[9-6], "With the aid of radio, rural inhabitants became more aware of the ebb and flow of life in the world, and it enabled them to play a less passive role in an industrial society."

Farm productivity began to increase in the U.S. in the 1930s (Figures 9-6 and 9-7). Several factors contributed to this growth, including farm machinery, tractors and low-cost fuels for their operation, electricity and water for irrigation from federal river development projects, and the use of lots of nitrogen fertilizer.

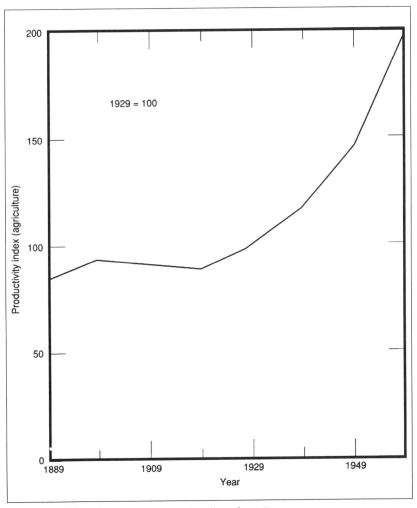

Figure 9-6. Historical trend in agricultural productivity.

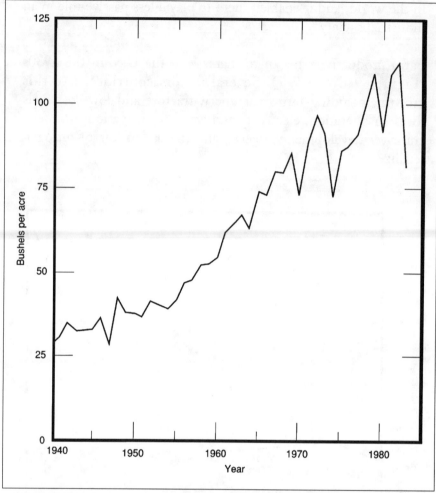

Figure 9-7. Trend in corn yield.

Irrigation

Sprinkler irrigation techniques reduce water use by 30 to 50 percent but require about ten to twenty times as much energy to distribute water as does field flooding. Total irrigation acreage increased by 5.5 million (10.5 percent) from 1976 to 1984, while acreage using pressurized irrigation techniques increased some 7 million (48 percent). These increases reflect new irrigation practices in naturally watered areas that never used flood irrigation as well as conversion of flood-irrigated systems to sprinkler irrigation in water-starved Western states (Table 9-3).

The benefits of pressurized irrigation systems must go far beyond those associated with surface flooding to justify their extra capital cost and energy use. Besides saving water, the pressurized system can deliver precisely metered quantities of water and chemicals uniformly to the growing crop. The control available in this way reduces the fertilizer requirement by 30 to 40 percent. Uniform water application permits the farmer to maintain a relatively constant moisture level in the soil, which often results in a more consistent, higher quality crop. Without surface flooding, some field operations can proceed even as the crop is receiving sprinkler water. One person can manage ten times as much sprinkler-irrigated acreage as surface-irrigated acreage.

Table 9-3
Change in Total Irrigation Acreage
and in Sprinkler Irrigation Acreage
in Western Regions, 1976 to 1984

Region	States	Change in Irrigated Area			
		1000s of Acres		Percent	
		Total	Sprinkler	Total	Sprinkler
N. Plains	KS,NE,ND,SD	+3356	+2752	+37	+101
S. Plains	OK, TX	−615	+262	−7	+11
Mountain[1]	CO,ID,MT,NM, UT,WY	+454	+659	+3	+22
Pacific	CA,OR,WA	+1337	+1604	+11	+49

[1] Data incomplete for Arizona and Nevada.

The trend toward increased irrigation during crop growth periods will require more electricity. Current increases in acreage under sprinkler systems are occurring in states where the irrigation systems have either traditionally used electricity or are converting to electricity (from gasoline and diesel fuel) (Tables 9-4 and 9-5).

Table 9-4
Patterns of Energy Use for Irrigation in 1974

Region	States	Source of Irrigation Energy, Percent of Total		
		Elec.	Gas[1]	Other
No. Plains	ND,SD,KS,NE	19	41	40
So. Plains	TX,OK	17	75	8
Mountain	ID,MT,WY,AZ,NV, UT,CO,NM	57	35	8
Pacific	WA,OR,CA	99	1	

[1] Natural and liquefied petroleum gases.

In states such as Washington, where agriculture depends on electric-powered irrigation systems, electricity provides most of the energy used for agricultural production. In 1972, over three-fourths of all energy used by agriculture in Washington was derived from electricity (Table 9-5). Less than 30 percent of the irrigation acreage used sprinklers that year. By 1984, 70 percent of Washington's irrigation acreage used sprinklers, which implies that the electric share of Washington's agricultural energy had grown to about 85 percent.

Table 9-5
Energy Input and Electricity's Share of Direct Agriculture Energy Use in Washington State, 1972

Wheat		Other Field and Seed		Fruits and Vegetables		Livestock		Total	
On-Farm Energy Use 10^{12} Btu	Elec. Share %	On-Farm Energy Use 10^{12} Btu	Elec. Share %	On-Farm Energy Use 10^{12} Btu	Elec. Share %	On-Farm Energy Use 10^{12} Btu	Elec. Share %	On-Farm Energy Use 10^{12} Btu	Elec. Share %
9.8	57	4.3	58	14.0	89	14.2	87	42.3	78

Notes: Excludes energy embodied in fertilizer and chemicals; electricity counted at 10,600 Btu per kwh.

Advances in irrigation technology contribute to electricity's growth in the agricultural sector. Commercial center pivot irrigation systems now serve an entire section of land (640 acres) from a single pumping point. The irrigation towers, spaced nearly 200 feet apart and kept in alignment by laser beam, move in a circle driven by individual electric motors. In another example, high dosage ultraviolet radiation successfully destroys the DNA in iron bacteria. Iron bacteria feed on the iron content of the irrigation water, and, because of their rapid multiplication rate, cause drip irrigation systems to clog. Electrostatic charging of pesticide droplets uses pesticides three to four times more efficiently than conventional spray techniques.

In the future, farmers could adapt an electrical distribution system used for irrigation to serve vehicular motors for field operations—especially if combined with no-tillage cultural practices. Irrigation and no-tillage culture are synergistic in that no-tillage practices conserve water. The combination of irrigation and no-tillage practices are now in commercial operation in Williston, Florida, and DeWitt, Nebraska.

Animal Husbandry

Direct energy consumption contributes only a small portion of production costs in animal raising enterprises; major items include feed, capital, and labor. Electricity is generally available at animal production facilities and it has played an important and expanding role in raising the productivity of animal production in the post-war period.

Poultry production offers an example. Today's poultry production is factory-like when compared with our great-grandmothers' efforts in their chicken coops. Flocks may number 100,000 birds. Poultry farmers confine them in shelters and carefully control and monitor climate, photoperiod, feeding, and sanitation. Ventilation fans, heaters, and evaporative coolers manage temperature and humidity. Mechanical feeders and waterers select the quantity and timing of feeding. For laying flocks, mechanization also controls the photoperiod and collects, grades, and packs the eggs. Electric systems power all of these environmental control operations.

Electricity provides all the energy for many animal raising needs such as lighting, ventilation, milking, and milk cooling. Even for space and water heating, farmers increasingly use electricity. In 1974, electricity met 60 percent of water heating requirements and 25 percent of space heating needs. In 1981, the numbers rose to 74 percent and 39 percent respectively.

New technologies contribute to the economic use of electric heat. In dairy operations, heat rejected from the refrigeration system that cools milk can heat water to wash the electric milking machines and associated milk-piping system. If an operation's water storage capacity is adequate, this heat recovery method can meet most of a farm's water heating requirement and use the rest of the electricity during off-peak periods.

With the shift from family-sized to much larger flocks, liquid petroleum (LP) gas largely displaced the use of electricity for chick brooder heat. More recently, the high cost of LP gas combined with new brooder house designs (that permit off-peak electricity use) signal a return to the use of electricity for this application. The building design incorporates a concrete floor with embedded electric heaters or pipes for hot water. The concrete floor stores and releases heat gradually so that the system can provide almost all of the electricity required during off-peak periods.

Electricity has not yet been used for farm vehicle operation. In the interim, researchers at South Dakota State University have devised electric chore tractors that operate on rechargeable lead acid batteries. Animal husbandry operations are an ideal application for the electric tractor because the tractors operate quietly and release no exhaust fumes thus, they can be used inside buildings and around animals. Nightly battery recharge allows owners to take advantage of low off-peak power rates.

The counterpart to the electric chore tractor in commercial activities is the electric forklift truck. More than half of all new forklift trucks now sold are electric despite their higher first cost and more limited performance. Their main economic advantages are high reliability and low maintenance costs. Once introduced, the electric chore tractor is likely to meet with a similar pattern of acceptance.

Manufacturing Activities Peripheral to Agriculture

Agricultural trends influence electricity use in manufacturing activities peripheral to agriculture. These activities include machinery manufacturing, petroleum refining, agricultural chemicals on the input side, and food processing on the output side. Electricity provides a growing share of the total requirement for energy in all agriculturally related manufacturing activities.

The substitution of agricultural chemicals for acreage harvested (e.g., use of nitrogen instead of harvesting more acreage) and for field operations (e.g., use of herbicides instead of cultivation in no-tillage practice) results in a net upward shift in electricity demand. This growth occurs because the electric share for chemicals production is greater than for petroleum refining (15 percent), the energy for chemicals (40 percent) exceeds the energy for machine operations (30 percent), and the production operations of both manufacturing activities are becoming more electrified. The net effect from 1974 to 1981 of extra electricity use per ton of fixed nitrogen (610 to 630 kWh/ton), extra nitrogen per acre harvested (57 to 68 lbs/acre), and extra acres harvested (321 to 361 million) was about 900 million kWh. Similarly, the electricity increase for refining the petroleum needed for all U.S. farming operations was about 200 million kWh.

Energy and electricity trends for food processing are given in Table 9-6 below. The data show that the sub-groups requiring refrigeration are the most highly electrified (meat, dairy, fruits, and vegetables). There are also the food categories that exhibit upward trends in consumer preference and are most highly electrified in their production.

Table 9-6
Energy Use for Food Processing in 1974 and 1981

Manufacturing Activity	1974 Purchased Energy, 10^{12} Btu	1974 Percent Electric	1981 Purchased Energy, 10^{12} Btu	1981 Percent Electric
Meat products	165	45	151	47
Dairy products	139	41	123	43
Fruits and vegetables	163	31	156	40
Grain mill products	189	31	205	37
Bakery products	73	38	68	38
Sugar products	164	12	142	15
Beverages	133	31	151	38
Miscellaneous	199	33	215	34
Total	1225	32	1211	36

Note: Electricity counted at 10,600 Btu per kWh.

In addition to the growing importance of refrigeration, technologies based on the use of semipermeable membranes and microwave heating also add to electricity used in food processing. Semipermeable membrane systems are displacing evaporation in food separations. For example, wort (the sugar solution) is now being separated in this way from spent grain in beer production. Other applications include concentration of whey (the liquid by-product of cheesemaking), skim milk, fruit juice, potato starch, and tea solutions.

Microwave heating in vacuum can remove a small amount of moisture from soybeans, which, in addition to drying the grain uniformly, "causes unwanted bean hulls to be released more quickly and completely and without overheating, than with conventional forced hot air systems." In contrast, hot air heating "yields a hot, nonuniformly dried product that must be held in tempering bins (where it unavoidably cools) for five to ten days to equalize moisture content" prior to being reheating for further processing.

Other microwave heating applications include proofing and frying doughnuts, thawing and tempering meat, and drying and cooking pasta. Proofing is the process of warming the yeast so that the dough will rise. Conventional air-heated proofers are hard to keep clean and control and require 30 to 40 minutes for proofing compared with four minutes in the microwave ovens. In meat tempering, meat is warmed from a hard frozen state to a temperature just below freezing where it can be further processed more easily and then refrozen without harm. With conventional tempering in warm rooms, temperature control is difficult. The time requirement can exceed eight hours for a 60 pound block of beef and can result in loss of as much as 15 percent of the meat juices. Microwave tempering cuts the time to five minutes—thus minimizing drip loss and bacterial contamination. Conventional pasta drying requires five hours in moist warm air to avoid skin hardening and later "checking" (cracking) as the trapped moisture breaks through the hard surface. Microwave units dry pasta evenly in 30 minutes under conditions that keep bacteria counts low.

References:

9-1 *Roles of Electricity: A Brief History of Electricity and the Geographic Distribution of Manufacturing*, Electric Power Research Institute, EM-5298-SR, July 1987.

9-2 *Roles of Electricity: Pulp and Papermaking*, Electric Power Research Institute, EU-3008-9-86,1986.

9-3 *Electricity in the American Economy: Agent of Technological Progress,* pages 136-138, Sam. H. Schurr, et. al., Greenwood Press, New York, 1990.

9-4 *Roles of Electricity: Glassmaking,* Electric Power Research Institute, EU-3010-7-86, 1986.

9-5 *Roles of Electricity: Petroleum Refining,* Electric Power Research Institute, EU-3011-7-86, 1986.

9-6 *Roles of Electricity: Agriculture,* Electric Power Research Institute, EU-3014-7-87, 1987.

CHAPTER 10
Electric Transportation (10-1)

The inadequacies of the U.S. transportation system grow daily. Increasing traffic congestion, insufficient public transit, and the noise, pollution, and resulting human frustrations both problems create are frequent sources of feature stories and commentaries. Virtually no person living or working in a major metropolitan area is untouched by this steadily worsening situation.

Today's mobility problems could become tomorrow's crises unless new transportation technologies solve these dilemmas. These technologies must relieve congestion on urban highways and increase accessibility to urban businesses, services, and activities. In addition, they should tackle the problems of air pollution and U.S. dependence on imported petroleum.

Reasons for Growing Transportation Crisis

Personal Travel

Personal travel in the U.S. has increased dramatically during the past three decades. Between 1960 and 1985, vehicle-miles of travel by road vehicles (cars, trucks, motorcycles, and buses) increased by nearly 150 percent—five times faster than the U.S. population.

The U.S. Department of Transportation's 1983–1984 National Personal Transportation Study gives a more complete picture of this dramatic growth in travel. This study is the most recent and complete effort to evaluate personal travel. Key findings, all of which compare travel data from 1969 with data from 1983, include the following:

- **Overall travel increased.** During 1983, people in the U.S. made more than 224 billion trips, traveling a total of almost 1,947 billion miles—55 percent more trips and 39 percent more miles than they traveled in 1969.

- **Number of trips increased.** The total number of households increased by 37 percent and the number of trips per household increased by 13 percent between 1969 and 1983.

- **Vehicle ownership increased.** Vehicle ownership increased by more than 40 percent between 1969 and 1983, from 0.7 to 1.0 vehicles per licensed driver. Greater accessibility to vehicles translates into more travel opportunities.

- **Non-work travel increased.** The number of trips for shopping, personal business, and family-related purposes increased by 39 percent on a per household basis. Trips per household for social and recreational purposes also increased.

In general, the greatest increases in travel have occurred in the discretionary trip categories such as shopping, social, and personal business. Researchers expect further increases in the volume of travel as the number of households increases and the population continues to grow and prosper.

Transit

Although the volume of travel is increasing, the share of travel served by transit is decreasing. Statistics show that transit ridership (in passenger trips) decreased by 16 percent, from 9.4 million in 1960 to 7.9 million in 1983. Nationally, in 1983, transit accounted for only 2.6 percent of all trips made.

Even the use of transit for travel to work is dropping. Data from 1983 show that transit's share of work trips was 6.2 percent. By comparison, 2.5 times as many work trips are made in light-duty trucks than for all modes of transit combined.

Movement of Goods

Urban and suburban transportation also moves goods. From 1960 to 1985, inter-city truck shipments, measured in ton-miles, grew by 110 percent. While this growth pertains to inter-city transport only, it does provide an indication of the vast growth in overall goods movement.

Trends That Shape Travel Demands

Making decisions about future transportation systems will require the consideration of key social, economic, and political factors that influence travel. These include:

- Demographic changes
- New land development patterns
- Governmental priorities
- Air quality and petroleum use concerns

The interactions between transportation demand and these factors are complex. The remainder of this chapter discusses the past, current, and future changes in these relationships. Recognizing and understanding these changes can illustrate the ways in which advanced electric transportation systems could help meet future travel needs.

Advanced electric technologies now under development can help alleviate transportation problems. In the future, these systems may satisfy the diverse demands of transporting people and goods and foster (rather than restrain) balanced regional development—without continually assaulting our environment. For example, many of these electric transportation technologies offer tremendous environmental benefits because of their potential for reducing CO_2 emissions. Electric vehicles (EVs) offer an important alternative for meeting the environmental challenges of today and the future.

Electric Vehicles: One Solution

For many years, the electric power industry has been working with major automotive manufacturers to develop reliable, practical EVs that meet industry and consumer standards. EV use could help mitigate two serious environmental problems: urban air pollution and global warming. Widespread EV use could contribute (albeit minimally) to increased power plant emissions of SO_2 and NO_X. However, use of emission reduction technology could help control these emissions with only minor increases in EV operating costs.

EPRI EV research efforts have produced two electric vans, the General Motors-sponsored Electric G-Van and the Chrysler-sponsored TEVan minivan. Both models are slated for use in commercial service fleets within the next few years. They were formally unveiled to the press and the public in 1992. Table 10-1 lists the technical specifications for each vehicle in the EPRI study. It compares the Electric G-Van with the operating characteristics of the GM gasoline powered van because they share similar load carrying features; likewise, the Electric TEVan is compared with the Chrysler minivan. EPRI's recent study indicates that electric vans (GVan and TEVan) release significantly less CO_2 into the atmosphere, when compared with similar gasoline powered vans at comparable driving conditions. The study classified emission standards based on three environmental perspectives: urban air quality, global warming, and acid rain.

Table 10-1
Vehicle Characteristics

	GM Gasoline Powered Van	Electric G-Van	Chrysler Gasoline Powered Minivan	Electric TEVan
Wheelbase	125 in	125 in	112 in	112 in
Gross Vehicle Weight	8600 lbs	8600 lbs	3975 lbs	5548 lbs
Cargo Volume	256 cu ft	256 cu ft	133 cu ft	133 cu ft
Range	—	60 miles*	—	120 miles**
Top Speed	—	52 mph*	—	65 mph**

* Simulated Urban Driving   ** Projected

For gasoline vehicles, sources of emissions include both direct (tailpipe) and indirect (those associated with electrical energy production, including mining operations, fuel transportation, energy conversion, and electricity transmission and distribution). Electric vehicles do not emit any pollutants during operation. Therefore, all emissions associated with EVs come from indirect sources—the electric power plants that provide electricity to the vehicles.

The reduction in CO_2 emission depends on the method used to generate the electricity that powers the vans. In the case of a TEVan powered by the national electric generation mix, CO_2 levels were reduced by 55 percent. If the TEVan is powered solely by electricity generated from coal plants, CO_2 levels dropped 20 percent. For this study, EPRI provided EV emissions from different present and future generation scenarios to bracket the range of possibilities. Emissions levels should drop further in the future, because new fossil fuel plants will be considerably cleaner than those operating today.

Table 10-2 shows the energy efficiencies of current gasoline and electric vans. These values assume that both types of vans operate as fleet vehicles in an urban environment. For electric vans, the energy consumption value (Btu/mile) has been adjusted to include indirect emissions.

Table 10-2
Energy Efficiencies of Gasoline and Electric Vans

	GM Gasoline Powered Van	Electric G-Van	Chrysler Gasoline Powered Minivan	Electric TEVan
Fuel Efficiency	10 mi/gal	1 mi/kWh	16 mi/gal	2 mi/kWh
Energy Consumption	14,400 Btu/mi	10,800 Btu/mi	9000 Btu/mi	5400 Btu/mi

Table 10-3 compares gasoline powered van emissions based on current California and U.S. 1989 emission standards with EV emissions under different generation scenarios. Table 10-4 compares emissions of CO_2—the principal emissions linked with global warming—for gasoline and electric powered vans. Table 10-5 compares SO_2 and NO_X emissions from gasoline and electric powered vans. The comparisons between gasoline and electric powered vans suggest that EV use could help mitigate two serious environmental problems: urban air pollution and global warming.

Table 10-3
Fleet Van Emissions Associated with Urban Air Quality
(grams per mile)

	GM Gasoline Powered Van			Chrysler Gasoline Powered Minivan		
Current Emissions	VOCs	NO_X	CO	VOCs	NO_X	CO
California	0.8	1.1	0.02	0.7	1.1	9.0
U.S.	1.1	1.8	10.0	1.0	1.8	10.0
	Electric G-Van			Electric TEVan		
Generation Scenarios	VOCs	NO_X	CO	VOCs	NO_X	CO
LA Basin Current	0.02	0.17	0.02	0.01	0.08	0.01
U.S. Current	0.02	2.5	0.1	0.01	1.2	0.05
U.S. New Post–1995	0.02	0.7	0.1	0.01	0.3	0.05

Table 10-4
Fleet Van Emissions Associated with Global Warming
(grams per mile)

	GM Gasoline Powered Van	Chrysler Gasoline Powered Minivan
Current Emissions	CO_2	CO_2
U.S. Average	1100	690
	Electric G-Van	Electric TEVan
Generation Scenarios CO_2	CO_2	
Calif. + Import Current 390	195	
U.S. Current	630	315
U.S. New Post–1995	640	320
Coal Plant Current	1090	545
New Coal Plant Post–1995	1030	515

Table 10-5
Fleet Van Emissions Associated with Acid Rain
(grams per mile)

	GM Gasoline Powered Van		Chrysler Gasoline Powered Minivan	
Current Emissions	SO_2	NO_X	SO_2	NO_X
California	0.2	1.1	0.2	1.1
U.S.	0.2	1.8	0.2	1.8
	Electric G-Van		Electric TEVan	
Generation Scenarios CO_2	SO_2	NO_X	SO_2	NO_X
Calif. + Import Current 390	0.7	0.8	0.3	0.4
U.S. Current	6.0	2.5	3.0	1.2
U.S. New Post–1995	1.0	0.7	0.5	0.3
Coal Plant Current	10.0	3.6	5.0	1.8
New Coal Plant Post–1995	2.0	1.0	1.0	0.5

Transportation Tomorrow

In addition to mitigating the environmental impacts of today's transportation systems, the U.S. must also address the problem of continually increasing traffic volume. Stop-gap measures to reduce traffic congestion will inadequately address the problem; instead, planners must introduce new transportation technologies and strategies to control traffic volume.

Future transportation systems, whether electrically powered or not, must work with and complement systems already in place. In addition, they should:

- Respond to peoples' demands for personal mobility and the use of private vehicles
- Provide mobility for those who cannot drive private vehicles, such as the young, the elderly, the handicapped, and those who do not own a vehicle for transportation
- Meet consistently the changing demographic trends, land development patterns, and governmental priorities
- Offer environmentally acceptable solutions

Advanced electric transportation systems satisfy these objectives and could potentially play a major role in the future transportation. Two examples are electrified roadways and personal rapid transit.

Electrified Roadways: A Private Transportation Option

Electrified roadways would allow individual travel in privately owned vehicles and offer complete flexibility in routing a trip. An electrified roadway makes it possible to drive a dual-mode electric vehicle (DMEV) long distances without stopping to recharge the propulsion battery.

The two key elements of this system are the electric power source embedded in the roadway and the DMEV, which is equipped with an electric powertrain, a roadway power pickup, and a separate onboard battery system to store energy.

Personal Rapid Transit: A Public Transportation Option

Personal rapid transit (PRT) is a public transit system that could provide a high degree of mobility to persons who do not have private vehicles. PRT would use small automated vehicles to provide individualized public transportation. PRT cars, which carry up to four passengers, would travel on dedicated guideways between user-selected origins and destinations. A chief advantage of this system over other transit systems: users could route PRT cars flexibly and optimize them for each user's particular start and end points on the system.

Off-line stations and vehicle-control automation would maximize the system's speed and safety. In addition, PRT would allow considerable freedom in placing and routing guideways, since vehicle and roadway space requirements would be minimal. And, by increasing or decreasing the number of cars on the system, service could be tailored to demand. As an alternative to personal travel in privately owned vehicles, PRT could help reduce traffic congestion and thus improve traffic flow. And since electricity would power PRT, this system would improve air quality and lower the amount of petroleum used in transportation.

Implementing New Electric Transportation Systems

Introducing advanced electric transportation systems will first require broad public awareness of and support for electric transportation options. As the U.S. increases the use of electric technologies within existing transportation systems it will move toward accomplishing these dual requirements.

Transportation planners should place early emphasis on expanding the use of commercially proven electric transit systems such as light rail, people movers, commuter rail, and electric buses. These systems offer significant energy efficiency advantages (in addition to mitigating environmental impacts). Calculations show that the next generation of electric fleet vans will have a 60 percent efficiency advantage over their gasoline counterparts. Electric buses are about 85 percent more efficient per passenger mile than diesel-powered buses, and shipping freight by electrically-driven trains is about 45 percent more efficient per ton than shipping by semitrailer truck.

Currently, many cities rely on these systems to move commuters, shoppers, and sightseers around. In addition, they should push the use of state-of-the-art electric technologies (energy storage systems, power electronics, and process control systems) to make transit more efficient and economical. These activities will help build solid working relationships with transportation agencies and will provide valuable experience in transportation operations. Development of the most promising advanced electric transportation systems (e.g., electrified roadways and personal rapid transit lines) will proceed concurrently.

As the advanced systems become available, they will be tried on a small, introductory scale. For example, a demonstration of electrified roadways is planned in Santa Barbara, California. The plans call for electrifying a segment of State Street, the main road through downtown Santa Barbara, for use by a fleet of dual-mode electric buses.

New communities and those undergoing substantial renovation provide fertile grounds for testing the suitability of advanced electric transit systems. As traffic congestion and land costs increase, land developers now recognize that they must incorporate transit planning into a development's infrastructure and feasibility plans.

Advanced electric transit systems can also meet the transit needs of areas with environmentally or economically based restrictions on road development. As an alternative to expanding the roadway system, the use of electric transit would reduce the amount of space needed for roadways and parking and contribute to a cleaner living, working, and shopping environment. For example, in Las Colinas, Texas, a new business and residential development near Dallas, an automated people-mover system serves this new community's urban center and its adjacent residential area. Owners of land parcels serviced by the system are responsible for financing the segment of guideway transversing their property. The system is free of charge to users and operating funds come from property assessments.

Working with developers, transportation agencies, and regional planners, electric utility representatives can identify joint land/transportation development ideas that are compatible with environmental concerns and regional growth objectives.

Reaping The Rewards

Advanced electric transportation systems offer both community and nationwide benefits:

- Reduced traffic congestion through increased road capacity.
- Increased efficiency of transportation service through increased travel speed and reduced service cost.
- Improved safety through vehicle automation and technology innovation.
- Improved air quality and balanced use of domestic energy resources through substitution of electrical for petroleum-based power.

In addition, these systems offer the electric utility industry an opportunity to greatly increase the market for electricity.

Reaping these rewards will depend on the electric utility industry's willingness to support and promote the development of advanced electric transportation systems. EPRI, working with electric utilities, government agencies, and transportation authorities, is developing a systematic and balanced approach to building a future transportation system that integrates advanced electric transportation concepts with existing roadways and transit operations. From this foundation, it will be possible to develop transportation systems that can accommodate future growth and achieve a high level of transportation service and personal mobility.

References:

10-1 *Advanced Electric Transportation Systems: Meeting the Needs of the Future*, Electric Power Research Institute, EU-3017, 1988.

SECTION IV.

Strategies For the Future

CHAPTER 11

Customers' Needs

Electric utilities provide energy services. The better they do this job, the more clearly they convey electricity's value to consumers. Accordingly, the closer the match between actual consumer needs and wants with energy service offerings (products, programs, services, and end-use technologies), the more consumers will want to purchase utilities' services. And, once making the purchase, the more they will value these services. This chapter focuses on how utilities can better understand customers' needs. The next chapter further elaborates on the concept of "value."

Customer Needs: Basic and Derived

Human beings have some very basic needs, such as food, shelter, good health, and a wish to belong in a group. These basic needs, listed in column A of Table 11-1, ensure survival. When an individual cannot fulfill these needs at the desired level, he/she will act to meet them.

Electric utility customers' demands for energy and for energy consuming devices derive from these basic needs. An illustrative list of the derived needs fulfilled by energy products/services appears in column B of Table 11-1. By fulfilling a derived need, a person can partially meet one or more basic needs. For example, heated water washes soil, bacteria, and perspiration from clothing (via a washing machine) and/or one's body (via a shower). Thus:

- Hot water keeps one's clothes and body looking and smelling clean. Cleanliness (a derived need) helps fulfill a basic need to belong in a group, since most people want their companions to be reasonably groomed.

Table 11-1
Basic and Derived Needs

A Basic Needs	B Energy Product/Service Needs
Nutrition (food and water)	Warmed/cooled living space Cleanliness Cooked food
Air, water	Hot water Washing machines
Protection from environment (comfort and security)	Light Space conditioning Ventilation Burglar alarms
Belongingness (includes communication ability) and accessibility (transportation and communication)	Telephones, TV's, etc.
Work/production (process, heat, machine drive, etc.) Sense of worth/individuality (identify, choice/control over one's life)	Industrial processes Computer technology
Health (freedom from severely debilitating mental or physical condition)	Medical diagnostic tools

- Hot water helps remove disease causing bacteria from one's clothes and body. Decreasing the chances of disease helps maintain health (a basic need). Heated water also helps prevent chills by reducing the body's thermoregulatory stress, and thus is also conducive to health and comfort.

Basic needs occur constantly through time; derived needs, on the other hand, depend on the individuals involved. For example, on a hot summer day, people take different steps to protect themselves from the effects of heat (a basic need). Some people prefer hiding out in air conditioned spaces, some use electric fans, some spend lazy afternoons and evenings relaxing on open-air porches, and some find a big shade tree and rest under it.

Similarly, different people (or the same person in different years) may meet a derived need such as hot water in several ways. For example, a gas water heater, an electric water heater, or a solar-powered water heating system can each fulfill this need. This example illustrates how an energy product/service can fulfill basic (health, cleanliness) as well as derived (comfort) needs. Accordingly, this particular product/service has multiple characteristics or attributes.

Each individual prioritizes commonly shared needs differently—thus, the provision of energy service depends on factors such as demography, attitudes, beliefs, and a complex web of outside influences.[11-1] In addition, each individual's priorities change through time—energy purchase decisions, for example, depend on income, geography (and the resulting climatic conditions of a particular area), and trends.

Utility Customer Expectations: Findings From a Recent Study

To fulfill basic and derived needs, customers expect certain things from their electric utilities. First, they expect clean, reliable, adequate power. Second, they want low bills. Over the years, customers have come to assume they will receive these elements. In addition, they want to perceive that their utility is attentive, responsive, flexible, courteous, caring, and a good corporate and community citizen. They expect their utility to manage costs wisely, communicate with clarity and content, and maintain an effective service support infrastructure. Finally, increasing numbers of customers demand service options that fit their needs.

When customer expectations do not match utility actions, then utilities can face a variety of problems, including unhappy customers, increased competitive threats, and political pressures resulting from customers, regulators, and/or politicians.

How can utilities match customers' expectations? The first step is to learn how customers view utilities and their products. According to a recent Edison Electric Institute (EEI) study, residential customers think about and evaluate electricity in two ways.[11-2] They perceive that electric utilities provide both a product and a service. When considering electricity as a product to purchase, they evaluate its costs, its value to them, and its controllability. Customers also perceive that electric utilities provide a service; this perception matches the more traditional notions of service reliability and power quality. They measure the level and quality of this service based on the utility's interactions with them and their neighbors. From these interactions, they can then judge how much their utility cares about them and values their business.

EEI's study demonstrates that customers base evaluations of their electric utilities on trade-offs. They will tolerate one perceived "negative" (e.g., high rates) if they feel they get something else of value (e.g., high quality service). In fact, customers make tradeoffs between positive and negative perceptions of all businesses. According to EEI, because customers consider two factors (product and service) then utilities can position their interactions with customers in two ways—"as a superior product provider, and as a superior service institution that cares about meeting its customers' needs."

Research shows that customers assess the success of utility programs and services based on factors such as reliability, educational programs, bill payment procedures, cost control incentives, and community relations activities. Customers rank two offerings highest: bill payment programs and cost control incentives. If they perceive a utility shows genuine concern in these two areas, they are more likely to express positive perceptions about their utility. These two programs are followed in importance by activities that highlight customer involvement, communications activities, educational programs, and community relations.

Factors Affecting Customer Behavior

Basic customer needs and expectations often drive behavior patterns. Certain factors, such as demography, attitudes, and beliefs also contribute to customer behavioral patterns.

Demography

Demographers study changes in vital and social statistics (e.g., births, marriages, deaths, patterns of migration); they interpret these statistics to characterize past human behavior and project future activities. Demographic changes often unfold gradually; however, as they occur they change markets and modify society. For example, three major demographic changes are shaping today's residential customer market: (1) family units continue to diversify; (2) women's roles evolve, and (3) a post-World War II baby boom has changed the breakdown of customer age groups.[11-1] In addition, the baby "dearth" of the 1970s and the consequent "boom" of the late 1980s and early 1990s will also strongly impact the demographic breakdown of future residential customer markets.

First, the traditional nuclear family unit, comprised of a married couple with two children, continues to decline in the U.S. Instead, "family units" are growing smaller and diversifying. The number of households with a married couple and children has dropped from 40 percent in 1970 to 28 percent in 1985.(11-1) Instead, childless couples, blended families, single parent households, and an increasing number of divorced, unmarried, and elderly people who live alone comprise growing percentages of the population. Demographers predict that through 1995, most of the increase in new households will come from these "non-nuclear" families, which will, in turn, exhibit different energy needs, preferences, and behaviors.

The movement of women into the work force presages one of the largest economic and social shifts of the twentieth century. Economically, working women contribute dollars to budgets—either their family's or their own—and more of this money purchases services such as cleaning, laundry, childcare, and pre-prepared food (from both restaurants and the frozen food section of the supermarket). Since households will continue to reconfigure the division of labor, service needs will broaden as families trade dollars for time, and buying habits will become more convenience-oriented in the future.(11-1)

Socially, women now support themselves with greater regularity and fewer fit the traditional role of "homemaker." Accordingly, the loss of this full-time person to run the household has created intense pressures within marriages and families—time is becoming an increasingly precious resource. The implication for businesses, including the energy service business, is that customers' scarcest resource is no longer *money*; it is *time*.

The Post-World War II baby boom produced a huge group of children that have grown up and started families of their own. Predictions show the American household will average only 2.3 people in the 1990s, down from today's average of 2.7. Much of this decline results from lower fertility rates. In the early 1970s, U.S. couples had an average of 2.1 children; today, this average has leveled off at 1.8 children per couple. In some Western European countries, the rate is as low as 1.3 children per couple.(11-1) While falling fertility rates contribute to part of this decline, changing attitudes also play a role. As people become economically independent, they increasingly express a desire for individual privacy. To gain this privacy, they create new households of their own. Electric utilities must adjust their planning processes (as well as their product/service mixes) to serve these new configurations. If they fail, then customer satisfaction will decline.

Attitudes and Beliefs

Attitudes and beliefs shape a society's behavior. For example, after WWII, U.S. society adopted a series of assumptions that can be summarized as "if we play by the rules of the game we will be justly rewarded." Many members of this post-war generation still strongly hold these views. However, events (and the resulting political and social changes they incurred) during the past twenty-five years have altered this assumption for the children and grandchildren of this generation. Recognizing the assumptions a society makes—its attitudes and beliefs—as well as how these assumptions evolve and eventually erode helps us understand shifting American needs and wants.(11-3)

For example, Alvin Toffler advances two contrasting images of society's potential evolution.(11-4) The first image he presents assumes a world essentially as we know it today (same economic framework and political structure). The second image postulates that examining today's society will not begin to indicate the structure of tomorrow's; he suggests that, in this scenario, the future will dramatically differ.

On the other hand, Ann Clurman, Senior Vice President of Yankelovich, Skelly and White, feels that several important factors will affect future society:(11-3)

1. *The "new-collar" workers.* This mainstream group, twenty-five million strong, earns between $30,000 and $40,000 per year. They work as teachers, cab drivers, small contractors, and managers of small businesses. She considers this group particularly important because she foresees its values spread-

ing to the rest of the American population during the next two decades. "Members of this group dislike bureaucracy, greed, elitism, and they tend to be patriotic, committed to a sense of family and liberal on women's rights, abortion, and school prayer. They put a high premium on self-reliance and are concerned about a sense of control."

2. *The growing trend toward control and self-reliance during the next 15–20 years.* This trend will translate into the need for stability and contentment. People will show less regard for rapidly changing fashions and fads. Instead, they will become astute consumers; they will hold a new sense of consumer power and demand open and honest exchange of information to make one's own decisions. These new consumers will be educated and sophisticated shoppers who demand value and service and expect a more open and honest social contract between consumer and manufacturer.

3. *The development of an entrenched two-tier society within the United States.* On one hand, society has developed a group of "gourmet babies" in the past two decades; children who are given the best of everything and every intellectual advantage. On the other hand, one out of five babies is born out of wedlock in the U.S., many who will (and do) live in poverty and lack any real chance to develop intellectually. "We may be developing a permanent class of under-employed people and seeding conflict between haves and have-nots. Since we as a culture do not tolerate losers in large numbers, over the next 20 years we will be developing a new sense of working egalitarianism."

How do these predictions—Toffler's and Clurman's—relate to electric utility service? Utilities must study such predictions with their customers in mind and ascertain the impacts these factors may have on customer attitudes and behaviors. For example, if utility customers exhibit growing trends toward sophisticated shopping, then utilities must recognize this piece of market intelligence and make sure its products/services withstand the scrutiny of these customers. Consumers buy electricity to power many products. Electric utilities must look beyond the meter and beyond the outlet and focus on what electric energy does for customers, why they use it (or choose not to), and under what circumstances they make these choices.(11-5)

Residential Customer Preference and Behavior: What Motivates Residential Customers To Make Energy Purchases?

As customer needs, expectations, demographics, attitudes, and beliefs shift, traditional methods of estimating customer purchasing habits are becoming increasingly antiquated. Utilities that want to better predict how their customers purchase electricity and electricity consuming devices must learn more about customers' needs and the benefits they seek from utility products/services. Recent EPRI research has determined that residential consumers' energy needs fall into two categories: personal benefits and concerns.(11-6)

Consumers expressed four personal benefits they seek from energy service:

- *Convenience.* Residential consumers want convenience, such as that offered by microwave ovens.

- *High-tech enthusiasm.* Residential consumers want to own the latest technological equipment, such as personal computers, video cassette recorders, and compact disc players.

- *Comfort.* Consumers identify many different energy needs and benefits related to keeping homes comfortable (e.g., heating, cooling, or dehumidification).

- *Appearance.* Perceptions about appearance fall into two areas. Some consumers want to own the best available appliances because they offer superior performance and option choices. In addition, consumers believe that certain appliances enhance the aesthetic appearance of their homes and fit their chosen "lifestyles."

EPRI researchers also identified five consumer concerns that relate to electricity use and appliance purchases:

- *Electricity conservation/budgetary issues.* Consumers have strong opinions about the importance of conserving electricity, their own monthly electric costs (as well as other cost issues, such as appliance costs), and electricity's availability for future generations.

- *Personal control.* Residential consumers vary in their willingness to have restrictions placed on their energy use—some strongly want personal control, while others are willing to let the utility control their appliances.

- *Safety.* Not surprisingly, many are concerned about the relative safety of electric versus gas appliances (ranges and dryers), electric blankets, and microwave ovens.

- *Search minimization.* Some consumers are reluctant to "shop around" when replacing appliances; in addition, they dislike comparing appliance prices once an acceptable price has been found, calculating the operating costs of different appliances, and switching brands when replacing appliances.

- *Task versus area energy use.* Many consumers prefer task-oriented lighting (individual lights for different parts of a room) rather than area lighting (overhead lighting). This selection appears to reflect a general concern for task-specific energy use versus diffuse energy use. This interpretation is supported by additional qualitative and quantitative analyses which indicate a similar "task versus area" orientation in the use of energy for heating and cooling.

While most consumers share each of these nine benefits/concerns, they differ with respect to the *degree of emphasis* they place on each. EPRI's study concluded, however, that certain broadly defined groups of residential customers did share similar energy needs.

Needs-based Segments

EPRI researchers used market segmentation techniques to divide the broad, heterogeneous residential customer market into smaller, homogeneous markets. Based on the nine energy needs researchers discovered, they identified six distinct market segments. Interestingly, recent studies have shown that these six market segments—generally called needs-based segments—do not correspond to traditional demographic breakdowns used by utilities (e.g., rate class, end use, or energy consumption level). EPRI found that several needs-based segments share very similar demographic characteristics but exhibit diametrically opposed energy needs/benefits. These findings show that utilities must adopt needs/benefits segmentation schemes to analyze more accurately residential consumers' energy preferences. Needs-based segmentation also provides a more direct link between targeting and energy service program positioning.

Market segmentation is not particularly new to utilities; what is different is this use of energy needs as a method of segmentation. Traditional segmentation methods helped them better allocate costs for rate design and forecast energy use. For example, residential customers might be divided into groups based on type of dwelling, age, income level, or location of residence. Today, utilities are discovering that they can (and should) augment these traditional market segmentation methods with information about customers' primary energy needs. Needs-based segmentation helps utilities select programs customers want, allocate marketing resources more cost-effectively, and enhance utility program planning. Utilities can then tailor these programs to meet the needs of a particular segment (called a "target market") and then use the promotion and distribution channels preferred by these specific target markets. Studies conducted suggest that utilities can substantially increase customer participation if they target DSM programs at specific customer market(s).

Figure 11-1 summarizes the important characteristics found in each of the six residential needs-based segments and indicates each segment's share of the U.S. market (among electricity consumers in single family homes). Figure 11-2 illustrates the energy needs profile for each of the six residential segments.

- *Pleasure Seekers.* Pleasure Seekers present a uniformly positive needs/benefits profile. They are more concerned than other consumers with conservation and budget issues, personal control, convenience, high technology, comfort, safety, and appearance. Their energy use tends to be task-related, and they are not particularly concerned with search minimization. Overall, Pleasure Seekers have many energy needs.

- *Appearance Conscious.* Appearance Conscious consumers differ most markedly from other groups: they are less concerned about using energy in a task-specific way. These customers are also more concerned than others with appearance—both the image that new appliances present (e.g., owning the "newest" thing) as well as aesthetic benefits that new appliances can provide (e.g., house lighting systems that offer extra security and comfort). They are relatively more concerned than others about safety, but are less concerned about conservation and budget issues.

- *Resource Conservers.* Resource Conservers are more concerned than most other customers with conservation and budget issues, and are willing to relinquish control of their household energy use to conserve energy. Resource Conservers also tend to use energy in task-specific ways. This tendency correlates with an interest in certain convenience devices such as microwave ovens. These customers also appear relatively unconcerned with search minimization—they seem willing to expend search time to meet their needs.

Utility Marketing Strategies: Competition and the Economy 233

PLEASURE SEEKERS

- Like all the benefits: comfort, safety, appearance, personal control of energy use, convenience, and high-tech appliances
- Show cost concerns

APPEARANCE CONSCIOUS

- Exhibit appearance and safety concerns
- Less likely to monitor energy use

LIFESTYLE SIMPLIFIERS

- Show less concern about most needs/benefits: comfort, high-tech appliances, monitoring electric usage, personal control, convenience, or appearance of home
- Fall into low income student or renter category

RESOURCE CONSERVERS

- Exhibit concern about budget and environment
- Accept utility controls for cost savings

HASSLE AVOIDERS

- Minimize hassle in buying appliances
- Want personal control
- Worry the least about cost or safety

VALUE SEEKERS

- Invest time in buying appliances (comparison shop)
- Use the most conservation measures
- Exhibit limited concerns about safety or appearance

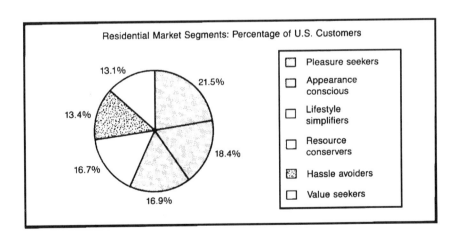

Figure 11-1. Residential energy needs-based segment descriptions.

234 Customers' Needs

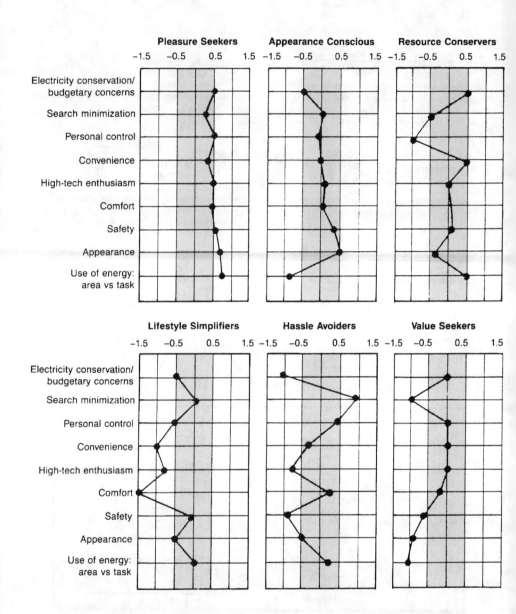

Figure 11-2. Relationship of the residential segments and their energy needs. The numbers on each grid are standardized factor scores— they indicate how positively or negatively a segment feels about each needs/benefits factor relative to the other segments.

- *Lifestyle Simplifiers.* Lifestyle Simplifiers exhibit the least concern with convenience. They also show less interest in comfort and new technology than any other group. However, they show average concern for safety, search minimization, and task-specific energy use.

- *Hassle Avoiders.* Hassle Avoiders want to avoid the effort and cost of searching for energy products/services. At the same time, they prefer to control their energy use personally. They are also willing to forgo savings, safety, and the benefits of high technology.

- *Value Seekers.* Value Seekers are willing to accept search costs in pursuit of other needs/benefits factors. They have a positive orientation toward high technology; however, they prefer less task-specific energy applications. Members of this segment are among the least interested in appearance and safety.

The Residential Customer of Tomorrow

When considering the future, electric utilities cannot precisely project how customers' needs will change. However, based on the past and present, they can draw a few conclusions. First, customers' energy purchase behaviors are not new—they have always been there. The industry's economic and regulatory environments simply allowed utilities to ignore descriptors such as needs, values, attitudes, and impacts of technological development. Utilities can no longer ignore these descriptors, which will remain valid for years to come—but their relative weights will change. While it is difficult to predict which attributes will decline, three will clearly become more dominant, as Figure 11-3 depicts.

1. *Economy:* The importance of energy cost and appliance efficiency will once again increase. Demands on income and general concerns regarding waste and conservation will increase as competition heightens and the percentage of

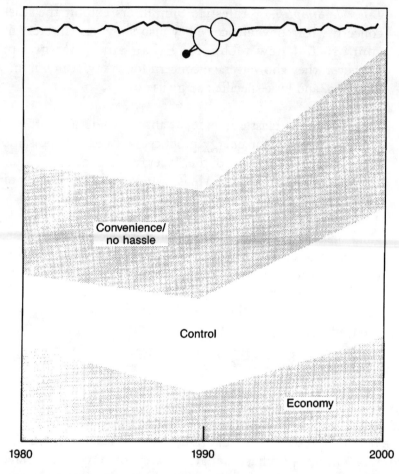

Figure 11-3. Some key customer needs are likely to change in the next decade.

   underemployed increases in the U.S.
2. *Control:* New consumerism will drive customers' demands for "more control." They will demand products tailored more specifically to their unique needs. This is not a return to Naderism, but a demand to "have the product I want."
3. *Convenience:* Pressures of time, changes in attitudes, the availability of technology, and continued increases in disposable income (for some Americans) are fostering a heightened importance of a product's convenience and potential to save time. The ability of energy service companies to respond to

these growing attributes may be the key to success in the future.

Commercial Customer Preference and Behavior: New Research Findings About Commercial Customers' Energy Needs

Electric utilities provide commercial consumers with a range of energy services, from artificial illumination to space conditioning; in addition, they power the increasingly complex equipment on which today's businesses rely. Figure 11-4 illustrates the potential growth of electricity sales in the commercial sector through the year 2000.[11-7]

Forecasters expect that, during the next 20 years, U.S. electricity sales to the commercial sector will grow by 2 percent to 3 percent per year. The fastest-growing regions in the country will continue to be the South and West, reflecting ongoing trends in population and employment growth. Factors influencing electricity sales growth include increases in: 1) new and retrofit construction activity, 2) space heating market share, and 3) the penetration of new electricity-using technologies. Factors which may limit electricity growth include: 1) increased appliance efficiency, 2) improved building designs which offer greater energy efficiency and utilization, 3) growing competition due to lower fuel prices, and 4) greater adoption of cogeneration applications.

Increasingly, the U.S. economy moves from being product- to service-based; this shift has resulted in dramatic changes in the commercial sector and its energy needs. The increase in service-based economic activity began with zeal during the Post-WWII period. Figure 11-5 illustrates the steady increases in electric intensity in kilowatt-hours per square foot of commercial buildings. This growth reflects electrification from space conditioning, water heating, electric cooking, and the development of systems to accommodate an information-based society. One consequence of the increase in commercial buildings' electricity use:

238 Customers' Needs

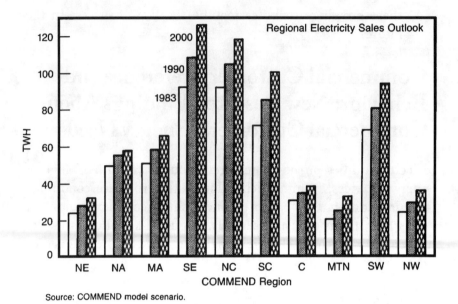

Figure 11-4. Electricity sales to the commercial sector (trillion watthours)

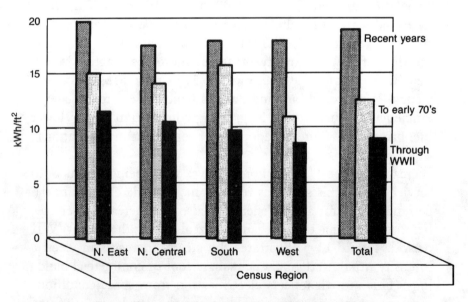

Figure 11-5. Regional electric intensity—commercial buildings.

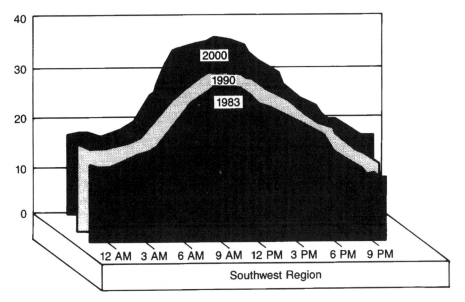

Figure 11-6. Commercial sector contribution to summer peak.

summer peak demands continue to grow (see Figure 11-6). Summer peaking utilities, in particular, want to improve commercial sector load factors to meet this challenge.(11-7)

Because of this sector's diverse range of businesses, sector decision makers need a wide range of energy products and services. However, market research suggests that they do share two very important energy-related needs: they want fairly high discount rates and relatively short payback periods for any energy improvement, retrofit, or service. These dual needs strongly influence electric utilities' designs for DSM programs as well as the incentives they must offer to make DSM programs successful.

Despite the changes in the commercial sector, many of today's office buildings (the largest commercial users of electricity) still look like those of fifty years ago—most offices still use the same basic desk and equipment layout. However, great potential for change exists, particularly when conducting space planning from an energy use and productivity perspective. For example, many businesses are finding that open planning system furniture offers increasing flexibility to match office layouts to job changes and position functions. These layouts also allow occupants to personalize their own spaces while still increasing the occupant density.(11-8) However, space planners can increase occupant density only so much. Studies show that workers need a certain amount of privacy or "space" to function at maximum productivity. This space allows the eye and mind to identify a sense of private place.

Electricity now plays an important role in establishing a new kind of office: the "electronic cottage." Futurists have been predicting this move from office-based to home-based workers for some time and technology advances (personal computers, modems) have hastened this change in work locations. In addition, the hassles of commuting on jammed freeways (and the resulting environmental impacts of too many cars on the road) and the need for flexible work hours have prompted many businesses and their workers to test this new work option. Futurists first envisioned the "electronic cottage" as a cozy room where a console of electronic gadgets would link the worker with his/her office.(11-8)

Electricity growth in the commercial sector is not a given: the natural gas industry has ambitious plans to counter electricity's inroads by developing and marketing technologies that compete with electricity. Both energy suppliers are targeting the heating market as the most viable and competitive area. Electric utilities have increased their shares sharply in recent years. For example, the electric heat share of older buildings is about 10 percent, compared to 40 percent for newer buildings. However, the gas industry's primary goal is to develop new technologies and restore the gas heating market share and, secondarily, to capture a larger slice of the cooling market. If it succeeds, it can reduce electric growth substantially.

Commercial Customer Energy Product/Service Preferences (11-9)

The commercial sector presents electric utilities with major opportunities and challenges. As with the residential sector, they cannot force customers to adopt new electrotechnologies and programs; customers must *choose* them. Many utilities have offered their commercial customers seemingly attractive programs and services only to be met with a lukewarm reaction. In fact, commercial sector participation in DSM programs has been relatively low—about 11 percent of the commercial businesses nationwide.

One reason for these "hit and miss" marketing efforts: the vast and varied commercial sector presents stiff challenges for DSM program marketers. Commercial businesses range from fast-food outlets and hotels to department stores and hospitals. Each company has its own culture, operating strategy, requirements for energy, and hardware. Furthermore, the decision to purchase a new piece of equipment or participate in a utility program is not made by a head of household, as in the residential sector; instead, the decision may be influenced by several individuals, each of whom considers the purchase on the basis of different needs.

This seemingly impenetrable complexity has hindered the development of effective tools for segmenting the commercial market. The traditional segmentation approach used by utilities involves segmenting commercial customers according to SIC code, building type, size, and mix of end-use equipment. This approach is useful, but it suffers from a fundamental weakness: it assumes that all customers within a particular segment will behave in the same way. For example, it projects that retail establishments with similar amounts of building space to cool and comparable peak electrical demand will respond much the same to a proposed space cooling program. Unfortunately, utilities have learned this assumption does not realistically occur in the real world.

Commercial customers who appear to be in similar businesses often have very different business strategies and operational needs. Consequently, their buying behaviors and program participation levels differ. For example, Bloomingdale's and K-Mart fall under the same SIC code and business type, but they operate with very different strategies, philosophies, and store layouts. Segmentation techniques that do not account for these strategic and operational differences are unlikely to produce good marketing results.

Many utilities have first-hand experience with this inability of traditional segmentation methods to meet businesses' needs. When attempting to get a commercial business to participate in a DSM program—a cool storage program, for example—the utility may win the support of the manager responsible for the company's energy operations. But the final decision to participate often comes from higher levels in the organization—the vice president of operations, perhaps, or even the chief executive officer. If the utility cannot convince these people a DSM program will meet the business's needs or strategic goals, they often veto the proposal.

EPRI's Commercial Customer Preference and Behavior Project offers utilities an entirely new way of looking at commercial businesses—an approach that takes into account the factors that determine *why* a customer will or will not participate in a DSM program. This new approach uses several tools, including an expanded version of the residential CLASSIFY™ model, called Commercial CLASSIFY PLUS, and a set of comprehensive questionnaires designed to obtain data from companies not only on their energy-related needs, but also on their overall strategy and operations. By providing insight into commercial customers' needs and purchase decision processes, this framework will help electric utilities target DSM programs at specific commercial customer segments as well as predict each segment's participation level in the programs. In addition to being a valuable tool for program planning, these EPRI products can help utility field representatives better understand their commercial customers' needs and thus serve them more effectively.

EPRI Segments The Commercial Customer Market

In developing its commercial sector project, EPRI conducted a series of interviews with businesses across the U.S. Each had recently decided whether or not to participate in a utility DSM program. The interviews revealed that in virtually every case, a diverse variety of needs contributed to each customer's decision. Customers were concerned about a DSM program's impact on business strategy, operations, and procedures. In addition, each questioned the utility's method of delivering the program and expressed reservations about specific end-use impacts. These concerns reflect business needs that utilities may or may not be filling. These basic needs are like rungs on a ladder: business strategy occupies the top rung, while business operations, energy operations, and end-use issues represent successively lower rungs. The bottom rung, called firmographics, consists of data such as SIC, building type, size, and mix of end-use equipment—the kinds of information that utilities now use to segment the commercial market.

When a company's needs are viewed in this manner, it becomes clear that the traditional approach—which focuses on the needs at the lower rungs of the ladder—provides a one-dimensional, bottom-up perspective that ignores the other levels of needs that influence a company's decision to participate in utility programs. These findings suggest that utilities must adopt a systematic top-down *and* bottom-up approach to identify needs if they want to adequately segment the commercial market.

Once EPRI researchers completed their initial series of interviews, they then talked with business strategy and operations consultants as well as managers and executives in the commercial sector. EPRI surveyed a broad sampling of two utilities'—Wisconsin Electric Power Company and Baltimore Gas & Electric Company—commercial customers. The survey respondents included managers responsible for their companies' energy-related needs as well as senior executives who could speak knowledgeably about their companies' business strategies and operations. Each participant answered a series of needs-based questions on a six-point agree/disagree scale.

EPRI then used the results from these interviews, meetings, and surveys to identify the dominant commercial customer needs that corresponded to each rung of the ladder. Researchers ranked these needs using a continuum that ran from strongly positive to strongly negative. Their findings: commercial businesses seek to satisfy a total of twenty-two primary needs when they are considering energy-related products and services. These twenty-two needs are distributed in a framework that matches the ladder.

For example, the business strategy level (or ladder rung) identifies a business's strategic needs. The next level, business operations, represents the needs that the chief financial officer or vice president of operations will seek to satisfy in managing the day-to-day administration of the company. The energy operations level represents the needs of the functional manager responsible for energy operations, while the end use level reflects the needs associated with specific end uses such as heating, cooling, and refrigeration.

Individual businesses vary considerably in the degree to which they value each need—their responses may range from a strong positive orientation to a strong negative orientation. One company may, for example, assign a high value to liquidity but a low value to long-range planning while another exhibits just the opposite orientation. Despite this diversity, commercial customers *do* fall into groupings that reflect unique patterns of needs.

Research has found that twenty-two needs exist in one form or another throughout commercial businesses, but there is considerable variation in the presence, intensity and direction of these needs among individual customers. Despite this finding, there are homogeneous groups of customers who place similar values on the twenty-two needs. Such groups can be found by applying a statistical procedure called cluster analysis to the twenty-two needs factors encompassed within the business strategy, business operations, and energy operations dimensions. In doing this, researchers identified nine commercial needs-based market segments. Each commercial segment is briefly discussed below.

Proactives

These organizations, which are very price competitive, have relatively high energy operation needs—especially for controlling energy use, establishing a proactive utility relationship, and obtaining sufficient/clean power. Their decision making and management style tends to be highly centralized, but with a long-range focus. Proactives have a strong need to invest in new technologies to maintain a competitive edge and, to a lesser extent, to realize gains in operating efficiency.

Besieged

Because of their weak cash flow position and the lack of long-range management objectives, the Besieged have few, if any, clearly expressed energy needs/requirements. Their focus seems to be dominated by the need to improve the company's financial condition.

Survivors

This group continually looks to improve cash flow at the expense of maintaining service/product quality. Survivors seek ways to reduce energy costs through special utility programs—customized services, improved rates, and billing flexibility. In addition, these businesses express a relatively high need to invest in new technologies to improve operating efficiency.

Innovators

Businesses in this segment are high risk takers, with a strong orientation to new product/service development—including a need for developing new technologies to maintain their competitive edge. Although their energy needs are not as strong as those of other segments, electric power is critical to their businesses, including the need for sufficient/clean power.

Utilitarians

With the exception of a focus on long range management, these service-oriented organizations have relatively few differentiated business strategy or operational needs. Utilitarians also express no strong energy or utility needs.

Dependents

Organizations in this segment tend to be conservative, expressing no strong interest in introducing new services/products or investing in technologies. In addition, they score relatively low on willingness to accept risk as a business strategy. However, energy is critical to their success, as reflected in their relatively strong need for backup generation, sufficient/clean power, and control of energy use. Consequently, Dependents seek a proactive relationship with utilities—including customized service and billing flexibility.

Conservatives

These service-oriented companies tend to adopt conservative business practices: they express a relatively low need for new technology and they are even less willing than others to take risks. Conservatives have a very centralized decision making process, but their management style allows for line responsibility to control costs. Sufficient/clean power is critical to their operations—but they have relatively low need for energy management, backup generation, or customized service. Additionally, their profits are perceived as being very dependent on energy rates.

Status Quos

Companies in this segment appear to be well-established businesses with a stable/mature product line. Status Quos express no strong energy or utility needs.

Self Reliants

These companies appear to offer quality products and services that are not price competitive. Specifically, they tend to score relatively high on emphasizing service/product quality before financials and low on price competitiveness. With this orientation, Self Reliants express low interest in new technologies as well as energy or utility assistance. Figure 11-7 illustrates the distribution of national commercial customers by these nine segments.

Each commercial segment exhibits a unique pattern of needs and thus presents utilities with different energy marketing opportunities and challenges. For example, Innovators, a segment representing about 8 percent of the commercial sector nationally, experiences slow and steady revenue growth, and as the label implies, segment members' business strategies are based

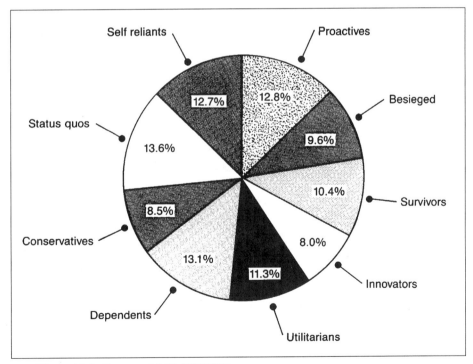

Figure 11-7. Commercial CLASSIFY Segments.

on constantly presenting new products and services. EPRI found that this commitment to innovation extends to their business and energy operations as well as their purchase behaviors. Innovators are generally driven to accept risks and adopt new technologies. In addition, their energy consumption is relatively low, yet they are committed to energy management. Figure 11-8 summarizes the attributes of this segment.

These types of insights into the needs and behavior of each customer segment can help utilities determine which ones will best respond to particular energy programs. For example, as described above, Innovators are generally interested in energy issues and will look to their utility as a source of advice on energy-related acquisitions. These are favorable characteristics as far as the utility is concerned, but they must be weighed against the small effect on utility loads that these customers would have, since they do not consume a lot of energy. Status Quos, by comparison, *do* consume high levels of energy, and thus their participation in DSM programs could have a large impact on utility loads. But these customers are generally indifferent to energy-

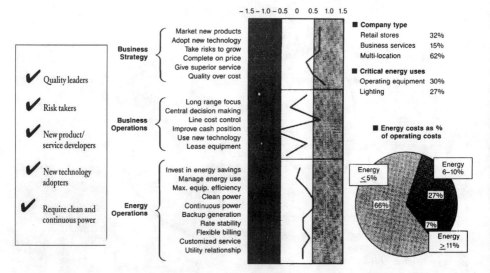

Figure 11-8. Innovators: attributes of this market segment.

related matters, so a utility would have to mount a sustained marketing effort to sign them up.

Using Commercial CP&B Findings

Utilities that wish to apply EPRI's findings to their own market segmentation efforts can follow a few simple steps. First, a utility must administer the segmentation questionnaires to a large sample of commercial customers in its service territory. Sometimes a single individual at a company—often the vice president of operations—can provide all the information required. In other cases, up to three individuals, representing different levels in the organization, may need to contribute data. Once a utility conducts the survey, staff members then feed the responses into the Commercial CLASSIFY PLUS program, which then divides the commercial market into the nine needs-based segments.

Based on the segmentation results, the utility then possesses a wealth of information about each commercial market segment. The model provides estimates of the percentage of customers in each segment and describes each segment's general business strategy. It details each segment's business and energy operations needs, firmographic and energy consumption profiles, decision making procedures, program participation rates, and promotion and distribution channel preferences.

Using these findings, the utility can analyze its commercial market at a level of depth and detail not attainable before. First, the utility can examine how its current and planned programs fit each segment's needs: will all, a few, or none of the segments participate? For example, Innovators are prime targets for cool storage and interruptible load programs, while the Dependents generally prefer efficient lighting programs. If offering a cool storage program, the utility might spend its marketing dollars more effectively by targeting the Innovators, as opposed to trying to reach the entire market. During this process, the utility may also identify *unmet* needs that may warrant the development of new programs or services. In addition to estimating program/service participation levels, Commercial CLASSIFY PLUS findings help utilities estimate each segment's load impact, by program, and then project the delivery cost to each segment. Combining these three pieces of information—customer participation, load impact, and cost analysis—helps utilities devise the optimal program/service portfolio for their commercial sector customers.

Adjusting The Marketing Mix

In addition to strategic applications, Commercial CLASSIFY PLUS helps utilities meet two important tactical objectives. The first is DSM program positioning: developing messages to inform key decision makers how the programs offered satisfy both utility and customer needs. The second is tailoring a program/service's marketing mix—product, price, promotion, and distribution—to make the package as attractive as possible to the largest number of customers.

Utilities face several challenges when positioning products/services in the commercial sector because business needs exist at several levels and purchase decisions are often influenced by several individuals. Traditional marketing approaches are often too simplistic because they direct positioning messages at the manager responsible for energy operations and ignore the fact that higher-level executives may veto a sale because the positioning message does not fit overall business needs. Commercial CLASSIFY PLUS, however, uses a multi-dimensional approach and permits a utility to develop separate messages for each decision maker. In addition, it identifies organization members who are likely to influence the decision and the type of positioning message that will encourage them to participate.

Case Studies: Products/Services Devised To Meet Customers' Needs

Electric utilities search continually for new ways to meet consumers' complex energy needs, because they recognize now that meeting these needs only enhances customers' perceptions of electricity's value. Utilities across the U.S. have adopted several tactics, including customizing service offerings and unbundling services. In addition, many utilities are discovering that issues such as reliability and power quality are important to customers. In fact, many of their customers are willing to pay a premium for higher quality service. Accordingly, they are creating services that focus on these two factors.

Customized Service Offerings

Electric utilities are beginning to customize service offerings. Their competitors and other product/service providers have aggressively adopted this tactic. Utilities are learning to apply their expertise in specific circumstances to offer entirely new products/services or add value to traditional ones. Unfortunately, matching service attributes to customer values is more difficult for an electric utility than for other energy suppliers because the needs of overall utility system operation require consideration. However, utilities are increasingly offering programs targeted at specific customer groups, such as thermal energy storage incentive programs for new construction (both residential and commercial).

Unbundled Services

Many utilities have begun to "unbundle" services, particularly for their industrial customers. These customers, in particular, can quantify their electricity needs for operations and processes as well as rate energy forms, in dollars, using factors such as precision, portability, speed, cleanliness, safety, variability, and energy content. For example, some industrial operations require less overall energy in electric form because of electricity's precision and/or efficiency in a particular application. Equally significant, other operations may need more electric energy but yield such savings in materials, time, labor, or environmental conditions that productivity overall is improved.

Electricity's Reliability

Consumers expect certain levels of reliability in each product they purchase. U.S. utilities provide an extremely reliable supply of electricity, however consumers may not consciously recognize this fact. Unfortunately, because of electricity's intangibility in most consumers' minds, they only think about it during infrequent power outages. Reliability is one of several attributes of electricity that utilities should emphasize when communicating electricity's value to customers.

Researchers from the Empire State Electric Energy Research Corporation (Eseerco) used focus groups and in-depth interviews to study utility consumers' perceptions of electricity's reliability. The study found that electricity consumers are generally satisfied with current levels of reliability; in most cases, they view outages as "acts of God," beyond utility control. In the study, consumers did express concerns about the frequency and length of outages. They ranked four or more outages a year or a single outage of more than eight hours as undesirable. Also, industrial customers expressed greater concern about the time an outage occurred (in terms of hour of the day, day of the week, or season of the year) than did residential customers.

A survey conducted for Pacific Gas and Electric Company (PG&E) examined the purchase decision consumers would make if presented with an explicit trade-off between electricity cost and reliability. Respondents were offered three program choices: a 10 percent reduction in electric rates in return for double the number of outages, a 10 percent rate increase in rates and half the number of outages, and retention of present rates and reliability of service.

The PG&E survey results varied greatly according to the type of customer. About 70 percent of residential customers indicated they would choose to decrease their monthly electric bill and accept lower reliability. Only 10 percent chose higher rates and reliability levels, and the remaining 20 percent preferred to retain their present level of service. Commercial and industrial customers overwhelmingly preferred the higher reliability option with its higher cost, even though these same customers had emphasized the need for lower rates on previous PG&E surveys. This preference for higher reliability reflects a strong sensitivity to the cost of outages and helps explain why the utility's previous DSM programs have attracted more residential than business participants.

The wide range of costs that may result from outages is particularly apparent from customer surveys conducted by Ontario Hydro[11-10] and the TVA.[11-11] For a twenty minute outage, the Ontario Hydro survey reported an average cost per kilowatt of load (in 1980 dollars) to be 4 cents for residential customers, $2.46 for large industries, and $6.72 for office buildings. Customers identified a wide range of preferred advance warning times (from a few minutes to 19.5 hours) to help them reduce costs.

The TVA market research study provides a more detailed analysis of how outages affect industrial customers. Among six industrial plants studied by TVA, the study found that the average minimum cost of a three hour interruption once a year was $324,000, while the cost of ten one hour interruptions per year was $396,000. The TVA study also found that total outage costs for industries in which mechanical drive was the major end use were from three to ten times higher than costs for industries that used electricity in furnaces or for electrolysis.

In general, studies show that, when faced with increasing curtailment of electricity, consumers prioritize their electricity needs and then cut back on those needs with the lowest ranking. For example, many first cut back severely on air conditioning and then reduce electricity use for water heating and miscellaneous loads. These same customers ranked refrigeration extremely high on their priority lists, because even when total electricity use was cut by 75 percent, these customers attempted to maintain refrigeration at virtually unchanged levels.

Power Quality

Because of customers' increasing needs for power quality, many electric utilities are seizing the opportunity to offer customers value-added services in the area of power quality. Utilities make every effort to ensure that the power they deliver meets strict standards set by the American National Standards Institute. However, sensitive electronic equipment can malfunction because of anomalies in the power delivered by utilities. Many of these customers purchase costly power conditioning or power protection systems in an attempt to correct the problem—an attempt that often fails.

Electric utility customers are installing computers, electronic cash registers, industrial process control equipment, facsimile machines, private telephone systems, home entertainment equipment, and other microprocessor-controlled equipment as never before. This equipment uses electricity, but unlike simpler electrical equipment such as motors, which tolerate greater variations in voltage, this sensitive electronic equipment requires a much more stable voltage source. Any variations in voltage—impulses, sags, surges, overvoltages, and undervoltages—can lead to malfunctions that range from minor processing upsets to total system shutdown. These malfunctions frustrate equipment owners and diminish the advantages of electronic technology.

Certain telltale symptoms tip off customers to power quality-related malfunctions in sensitive electronic equipment. These symptoms disrupt customer operations; furthermore, the symptoms may be difficult to reproduce when tracing the problem. For example, computers with power quality problems will often experience processing upsets: data on the computer screen may appear in wavy lines, or computers may suddenly stop functioning, or lose data, or experience disk drive crashes. Private branch exchange telephone systems may lose calls or exhibit high levels of background noise on the telephone line. Industrial process control equipment may experience sudden, inexplicable system halts. These represent only a few symptoms of power quality problems. In general, any sudden, apparently random occurrence is suspect.

New evidence suggests that utility customers should first check their own facilities carefully for the actual causes of electronic malfunctions before purchasing power conditioning equipment. Recent studies indicate that as much as 80 percent of all failures of sensitive electronic equipment attributed to poor quality power may result from inadequate electrical grounding or wiring on the customer's premises, or from interactions with other loads within the premises. This condition frequently can occur when installing electronic equipment that relies on existing building wiring. In many cases, proper grounding and wiring can correct the problem. After correcting grounding and wiring inadequacies, customers can more easily identify and rectify other problems. Utility customers should probably not consider installing power conditioning and power protection equipment until all other possible remedies have been thoroughly explored.

References:

11-1 "The World of 2006—Demographics and Lifestyle." *EPRI Journal*, March 1987.

11-2 *Residential Customer Competitive Positioning Study*, Edison Electric Institute by Cambridge Reports, Inc., 1988.

11-3 "The World of 2006—Values and Beliefs." *EPRI Journal*, March 1987.

11-4 Alvin Toffler. *The Third Wave*. New York: Bantam Books, Inc., 1981.

11-5 R. Whitaker, et al. "Is Cost the Only Measure of Electricity Value?" *EPRI Journal*, January/February 1989.

11-6 *Residential Customer Preference and Behavior: Market Segmentation Using CLASSIFY*. National Analysts, Synergic Resources Corporation, and QEI, Inc. Palo Alto, Calif.: Electric Power Research Institute, March 1989. EM-5908.

11-7 *COMMEND Planning System: National and Regional Data and Analysis.* Palo Alto, Calif.: Electric Power Research Institute, March 1986. EM-4486.

11-8 Tom Jarriel. "ABC AirReport." Produced for United Airlines by ABC News, ABC Video Enterprises, Inc., March 1988.

11-9 "Commercial Market Segmentation." *EPRI Journal*, 1990.

11-10 *Value of Service Reliability to Consumers.* Seminar proceedings prepared by Criterion, Inc., EA-4494, March 1986.

11-11 *Value-Based Utility Planning: Scoping Study.* Final report for RP-2381, prepared by Levy Associates and Meta Systems, EM-4839, December 1985.

CHAPTER 12

Service Marketing in the Electric Utility Industry: Next Steps

Introduction

Today's successful electric utilities serve their customers well. It is not an easy task. This chapter discusses how the concept and practice of providing service has evolved in the industry. It begins with a brief discussion of an element found in every successful service: value.

"I don't know what I want, but when I see it I'll know it." The concept of value is often amorphous to consumers; they cannot necessarily quantify it, particularly in its formative stages. However, once created, "value" is an extremely powerful service element. Just ask a much trusted family doctor—or Federal Express. Both entities offer services that customers value and will pay for, despite the cost.

Utilities (and regulators) get exceedingly uncomfortable when faced with factors that are difficult to quantify; concepts like "value" and "good service" certainly fall into this category. However, the way to add value (and thus enhance service) is no mystery: a utility increases customers' perceptions of value by adding services that customers want and need. For example, direct utility contact with customers is often key to gaining high acceptance rates of utility services. Customers appreciate the chance to talk with a knowledgeable utility representative. In ten air conditioner control programs, utilities that marketed intensively reaped a median participation rate of 70 percent, whereas those with minimal marketing achieved an adoption rate barely over 10 percent.

When a utility offers a service whose value to the customer *exceeds* its incremental cost, the utility takes a giant step toward meeting customer needs and improving its competitive stance. This focus on *value* relative to *price* gives utilities more options for increasing customer satisfaction. Today's successful electric utilities provide a wide range of services that customers value, including brokering cogenerated energy, making market-side investments, or concentrating on demand-side technologies and services.

Devising Service Strategies

Each utility must examine its own unique situation before selecting its service strategies. Often, utilities must first do some "detective work" to uncover market intelligence. A utility that has never functioned in a competitive environment may not know what its customers want, need, or will accept in the way of products and services. *A utility must know about its customers' purchase behaviors before it makes product or service decisions.* The utility also needs to more precisely identify and address the key competitive issues it faces in order to prioritize research needs, identify existing competitive research deficiencies, and learn how to open effective sales and service channels.

When devising its service strategies, a utility must identify several value-related factors. First, does the utility recognize the value (and importance) of spending scarce marketing dollars to better understand the customer? Customer perceptions of utility service directly impact utility revenues. If customers think they are not getting enough value (in the form of service) for their energy dollar, then they may do one of two things—seek the utility's competitor to meet energy service needs or become more dissatisfied. Conversely, if customers perceive they are getting an excellent value for their dollar, then they are less likely to either seek a competitor's service or become dissatisfied. In fact, these customers might raise utility revenues by increasing purchases of electricity service.

Second, does the utility feel it gets enough value for its dollar by increasing its sensitivity to customers' needs? Several marketing research tools can help utilities assess this point. They first compute the increased value customers perceive from new value-added services and examine the utility's cost of providing them. Then, these tools compare the findings.

Third, a utility must determine how to incorporate findings about value and good service into the DSM planning process. Traditional planning processes or newer ones like least-cost planning may not include a way to incorporate such a qualitative variable.

Fourth, a utility, in concert with its regulatory commission or elected board, must determine the impacts of using a "value of service" variable in program and rate development. In particular, is segmenting customers by homogeneous value of service "discriminatory"? If the programs and rates developed for customer segments bear no relation to cost of service, then a commission would probably respond negatively. However, different programs and rates should reflect actual variations in costs. Moreover, both commissions and utilities want satisfied customers. Thus, an effort to be sensitive to the customers' satisfaction or perceived "value of service" is in a regulatory commission's interest.

Since regulatory commissions have traditionally incorporated a "value of service" variable when setting rates for telephone service, utilities (and regulators) have a model to follow. Utilities may also find it useful to conduct additional research to analyze current methodologies for measuring value of service and articulating present day concerns. Perhaps such research should have joint sponsorship by utilities and regulatory commissions, as did the EPRI Rate Design Study on marginal cost issues.

Finally, the electric utility industry can further promote the concept of value by encouraging its members to adopt marketing programs. Marketing is the process of focusing on customers' wants and needs, providing products and services to meet them, and concurrently maintaining a healthy business. A good marketing plan can address a variety of strategies: it can encourage conservation or load management, help a utility build load, or maintain viability in a competitive environment. In fact, utility DSM activities form the basis of marketing—achieving mutual utility and customer benefit.

For years, utilities have responded to customer demand for electricity without clarifying exactly what service they provide. Customers, for their part, have consumed electricity and paid their bills at the end of the month without knowing exactly how they spent their money. This action is much like walking into a supermarket and purchasing items with no weights, measures, or prices marked on them. Only when the cash register tallies up the total bill does the customer know what he/she has spent. Utilities need to give customers more information—a better breakdown by end use—so they will have a greater sense of control over their electric bills. Customers value this sense of control; it contributes to developing an enduring, solid relationship between utilities and customers.

Understanding Customers' Needs

To be more successful in the increasingly complex and dynamic markets in which they operate, utilities are seeking ways to serve their customers better. This task requires doing two things well: listening to customers and cost-effectively delivering the products and services they want.

EPRI is developing concepts and tools utilities can use to assess and improve their commitment to customers. The Institute's integrated value-based planning project emphasizes customer-focused planning (CFP), which is defined as a *process* that makes the customer the focal point of *all* utility planning and activity and uses customer *value* as the driving force of business strategy.(12-1)

Key words in this definition highlight several important concepts. The first—indicated by the word *process*—is that, in spite of its name, CFP is more about *doing* than about *planning*. Customer needs are constantly changing, and to be successful, utilities must establish organizational procedures and processes that anticipate and respond to these rapidly moving targets.

The second concept, highlighted by the use of the word *all*, is that for this approach to be truly effective, the support for the concept of being customer-focused must permeate an entire organization—not just the marketing department, where its value is often best understood. The third, and perhaps most important, concept is that of *value*. The primary objective of CFP is to create value for customers through the continuous improvement of existing products and services and the development of new products and services that meet customer needs.

In terms of listening to customers, most utilities are typically in one of three stages:

- *Inattentive:* They neither seek nor hear customer opinions. For whatever reason, the utility is distracted.

- *Attentive:* They perceive the importance of meeting customer needs and accumulate customer information.

- *Seeking:* The utility and its customers work hand in hand to improve understanding and create value for customers and shareholders alike.

In terms of the ability to deliver services to customers, utilities are also typically in one of three stages of development:

- *Ineffective:* The utility may be listening, but its current levels of skills and resources prevent it from effectively meeting customer needs.

- *Reactive:* The utility responds to competitive threats and opportunities but succeeds primarily in holding its ground.

- *Proactive:* The utility is a leader in product and service quality and innovation and it works closely with customers and regulators to create and deliver value.

Figure 12-1 shows the stages of listening and delivering in a matrix format. The proactive, seeking utilities are clearly the industry leaders, while the inattentive, ineffective ones lag behind. At the other two extremes are the inattentive but proactive utilities and those that actively work with customers to understand their needs but cannot deliver services. The former are obvious—they are very good at what they do, but they are doing the wrong thing. The latter are obviously frustrated. In the middle are the followers, utilities that listen but respond only in a reactive manner and, at best, succeed in holding their ground relative to the competition.

Several of EPRI's research projects are designed to help utilities strengthen their commitment to customers. When a utility focuses on customers, it automatically serves its own objectives. Customers want to control costs and improve productivity. Both of these needs strongly drive utility efficiency. These needs also impact other customer objectives. For example, customers are becoming concerned about environmental issues. In part, this changing attitude reflects customer response to growing environmental regulations as well as the advantages of a "green" business strategy. Utilities also want (and need) to become sensitive to the environmental impacts of their businesses. Utilities can help customers recognize that electrotechnologies control costs and improve productivity and limit environmental impacts.

		Listening		
		Inattentive	Attentive	Seeking
Delivering	Ineffective	Lagging		Frustrated
	Reactive		Following	
	Proactive	Oblivious		Leading

Figure 12-1. Customer-focused planning.

The Introduction and Evolution of DSM Planning

DSM Definitions

The term demand-side management refers to a variety of methods for influencing customer demand, including energy efficiency, load building, and load shifting programs. These DSM alternatives also address another issue of increasing importance: they provide excellent vehicles for utilities to expand their energy service offerings. Since DSM programs seek to influence customer actions, utilities can tailor them to meet the needs of specific customer markets while accomplishing their own business related goals. For example, utilities are discovering that rebate programs offer an increasingly popular method of persuading customers to purchase more energy efficient equipment. About 35 to 50 percent of the nation's electric utility consumers are served by utilities that have some form of energy efficiency rebate program. Over sixty utilities currently offer rebates to promote high efficiency appliances such as super efficient air conditioners or other energy efficient equipment. The overwhelming majority (92 percent) of utilities offer rebates to purchasers to create market pull; about 24 percent of utilities provide rebates to appliance dealers to create market push. Rebates for high efficiency residential heat pumps, the appliance most commonly featured in rebate programs, range between $110 and $300 per unit.

Virtually all utilities in the U.S. now pursue DSM to some degree. In 1990, nearly 15 million residential customers participated in DSM programs. With healthy customer participation, DSM programs can potentially capture significant efficiency gains.

Projected DSM impacts accrue over and above the anticipated efficiency improvements that occur in the normal product evolution process. State and federal governments are providing one way of guaranteeing energy efficiency increases: they are establishing increasingly stringent state and federal appliance and building efficiency standards. In addition, as old, less efficient models wear out, customers who purchase new models automatically replace them with more efficient ones. Finally, as energy prices rise, consumers tend to buy more efficient technologies.

DSM History

The DSM framework was initially constructed for two reasons. First, it responded to the need for a logical process to help utilities optimize the supply–demand interface. Second, it presented a unique marketing tool. Most are surprised when the creator of the DSM framework reminds them that his intention was to sell supply-side planners on the basic idea that "demand need not be considered as fixed." It is almost comical that later, several leaders in the marketing revival of the late 1980s/early 1990s were critical of the concept. "Whoever invented DSM was an idiot" is one famous remark overheard by this author. (I had a burning desire to walk up and introduce myself as the idiot who did.)

DSM was, indeed, conceived as the premiere marketing ploy. It packaged the idea of planning both supply options and demand changes concurrently.

DSM was not the first marketing strategy utilities embraced. Edison himself used marketing representatives to promote electricity use and fill the daytime load valleys. As the industry evolved, this sales focus became a "market push" strategy. In this strategy, alternate pieces of hardware are pushed at or forced onto customers. Until the early 1970s, the strategy remained so narrowly focused that the only customer information utilities kept (aside from basic billing data) was the number of meters set.

The next step in the shift toward a customer focus was the introduction and evolution of load management. These initial efforts were hardware-centered and again, did not consider customer needs. The objectives were to identify technologies that worked in peak clipping, valley filling, and load shifting and push them into the marketplace. These load management efforts were followed closely by the early interest in "conservation." Again, technology was the driving force of these early conservation programs. Utilities maintained an inventory of all appealing programs; they would test hardware introduced by salespeople or that used by other utilities.

Instead of using a technology or hardware as the driving force behind these programs, DSM was the first marketing strategy that specifically promoted a customer focus. Instead of conducting "what if" studies on individual technologies, DSM allowed an integrated look at technology/customer/utility considerations. DSM grew out of the need for a convenient framework to think about the impacts of individual technologies and/or bundles of programs and initiatives on the marketplace.

Current U.S. DSM Activities

As we begin the 1990s, most utilities in the U.S. have initiated some kind of DSM program. Utilities with high reserve margins as well as those facing high marginal costs have come to recognize the viability of DSM alternatives. The most common DSM programs continue to focus on load reduction activities, however, increasing numbers of utilities (along with their regulators) are recognizing the potential benefits of load building activities, efficiency programs as well as those that enhance the supply–demand interface.

In 1991, the U.S. electric utility industry spent over $1.5 million on DSM programs. Some industry observers project annual expenditures of $5 billion by 1995 and $10 billion by 2000. Recent EPRI estimates suggest that DSM programs could reduce U.S. electricity use by 200 billion kilowatt-hours in the year 2000.[12-2] If favorable regulatory incentives continue to encourage DSM implementation, this number could increase to 450 billion kilowatt-hours by the year 2010 (an additional 11 percent reduction in electricity use).

DSM Dilemmas

DSM does pose a basic business dilemma. By focusing on demand-side services and products, utilities may spend money to reduce sales. If a utility faces a supply problem, then it has little choice. However, if it maintains adequate supplies and forecasts this trend to continue, it may hesitate to implement DSM. *Offering DSM programs/services does not necessarily mean settling for lower sales.*

Today, U.S. utilities and their regulators have devised incentive plans which allow recovery of DSM expenditures and, in many cases, incorporate a profit potential. The most comprehensive of these plans include DSM cost recovery, revenue loss recovery, and bonuses. As of late 1992, thirty-two states have adopted one or a combination of these elements, but the level and attractiveness of the incentives varies greatly from state to state.[12-3]

DSM *cost recovery* represents the most basic and oldest method utilities use to recover DSM costs. It allows utilities to recover all prudent expenses associated with implementing DSM programs (e.g., program planning, administration, promotion, market research, evaluation, and measurement). Utilities simply raise rates to recover the expenses. As of late 1992, thirty-two states allow DSM cost recovery.

Since DSM often results in lower kilowatt-hour sales and lost revenues, revenue *loss recovery* allows utilities to recover any loss of revenue. This element goes one step further than DSM cost recovery, which only recovers costs, not lost revenues. Revenue loss recovery removes one of the major disincentives utilities face when implementing DSM. As of late 1992, thirteen states allow utilities to recover lost revenue through rates.

The third element, *DSM bonus*, represents an actual incentive. In late 1992, nineteen states reward utilities for implementing DSM; utilities can actually *make money* in addition to recovering costs and lost revenues; six additional states offer partial incentives and at least five other states are currently considering implementing incentive programs. The specific incentive measures vary from state to state. Five basic types of incentives include: 1) a markup on expenditures, 2) a bonus return on DSM in the rate base, 3) a fixed bonus for each kW or kWh saved, 4) a utility's overall rate of return indexed to DSM performance, and 5) share of net savings in resource costs.

Many regulators and other industry influencers view these incentives with distaste. However, they discovered in the late 1980s that if utilities were to compete in an increasingly deregulated marketplace, they must be allowed one of the most basic rights of the free market: the ability to make a profit.

In addition, skeptics of regulatory reform feel this plan has one fundamental weakness: the difficulty of proving efficiency savings. They argue that increased regulatory involvement imposes "unnatural" solutions that further complicate the electricity demand–supply equation. In addition, they feel that not all utilities can afford to pass all cost increases on to their customers and electricity prices will eventually rise to the point where they are no longer competitive with alternatives.

Regulatory reforms alone will not improve energy efficiency. Free market activity must occur: utilities must see the potential profitability of entering the energy service business. Making money from energy services may sound similar to regulatory incentives, but the market-based solution would go much further. To some extent, the utility industry has begun to capitalize on the natural market for energy services. One example is Puget Sound Power & Light's establishment of a subsidiary called Puget Energy Services. The subsidiary focuses directly on selling and marketing energy services that encourage efficiency and load management. Its business includes leasing, renting, and financing energy-using equipment.

An Additional Planning Challenge: Regulated Marketing

Because utilities are still largely regulated businesses, they must plan and offer competitive programs and services that meet the approval of local, state, and federal regulatory agencies. When utilities faced little outside competition from other energy suppliers, their business strategies were relatively straightforward: sell electricity and keep the shareholders happy. Today, as competition becomes a reality—even in this "regulated" marketplace—utilities recognize that their business practices must change to meet this new challenge. However, many practices of "competitive" businesses do not meet the strictures of a regulated entity.

Currently, regulators and utilities are addressing this new business environment from a number of different planning approaches. There are two planning approaches of particular interest today: least-cost planning and integrated resource planning.

Today's utilities must devise integrated resource plans: those that consider DSM programs (including conservation and load management) as viable supply-side resources. In many cases, these programs cost less than constructing new supply-side resources. In addition, utility planners can include considerations such as risk minimization and available capital resources during a program's evaluation process. As with traditional resource planning methods, utilities adopting integrated resource planning still want to minimize costs to the extent possible. The two planning approaches introduced above, least-cost planning and integrated resource planning, offer different perspectives of cost minimization.

Least-Cost Planning

As utilities have increasingly adopted DSM strategies, many regulators and industry protagonists have stressed they follow a planning strategy known as "least-cost planning." Least-cost planning (LCP) seeks to create a mix of supply- and demand-side resources that will satisfy energy needs at the lowest possible cost to utilities and consumers within existing constraints (i.e., minimizing economic and environmental risks). It recommends that utilities consider both DSM (e.g., efficiency improvements, load management, and nonutility energy sources) and supply options when designing resource plans. In late 1992, at least forty-three states have adopted (by mandated policy or general practice), or are considering, LCP strategies.

LCP's primary objective is to improve the nation's energy efficiency. Under least-cost planning principles, utilities are increasingly permitted to earn a profit on load reduction programs. Utilities that adopt LCP generally focus on activities such as energy efficiency, energy efficient technology development, and improved DSM planning. In addition, LCP addresses environmental concerns such as global warming. However, most state regulators exclude load building options from the list of acceptable DSM activities.

The problem with the LCP concept is that it ignores two very important variables: market conditions and customer needs. As a result, utilities that use this planning concept (and the regulators that encourage it) may offer rebates and incentives to cut electricity use, but fail to consider the ultimate consequence of such an action. For example, when used incorrectly, a utility's rebate program might reduce electricity use in the short term but use more energy in the long term. If a utility offers commercial building owners incentives to switch from electric water-loop heat pumps to gas HVAC systems, they may save energy in the short term, but as the efficiency of electric systems continues to improve, these savings will disappear.

Industry protagonists who are particularly supportive of LCP are upset by DSM's inclusion of load growth strategies. However, they fail to recognize that the fundamental concept of matching present levels of supply and demand as well as optimizing future choices should allow planners to include growth and expansion of new generation capacity.

Integrated Resource Planning

In recent years, the LCP framework has evolved into one that stresses a rational balance of supply- and demand-side options. Planners have coined a new term—integrated resource planning (IRP)—when referring to this more pragmatic planning strategy. IRP incorporates economic parameters with those that can enhance electricity's value. The IRP process recognizes that cost is just one component of a customer's decision making process: some customers are more than willing to pay higher prices if they feel they are getting maximum value for their purchase. Thus, the IRP process produces programs that meet customers' actual energy needs more adequately and thus enhances their perceptions of electricity's value. In addition, under this process, utilities consider a whole range of resources—from traditional suppy to renewables, dispersed generation, energy efficiency, load management, and other DSM options.

Utilities that adopt the IRP process first make sure they understand their customers needs, values, and energy choices. Then, planners incorporate utility capabilities, limitations, and opportunities into the equation. Finally, they consider the regulatory environment and the marketplace as a whole. The conclusions result in the development of a set of program and service options designed to meet customer preferences more specifically. The options must include a wide variety of product variations; traditional and demand-side options will be considered, as well as a wide variety of other services and products. Obviously, utilities will find some options feasible to implement and some incompatible or unrealistic. However, this IRP process will result in program options that meet *both* customers' and utilities' complex needs. While utilities must incorporate information about customer needs and wants into an IRP process, they must also factor in their own goals into the equation. (Remember, an integrated resource planning process identifies and provides options that are valued by *customers*, attractive and feasible for *utilities*, and aligned with *public policies*.) A utility's goals can depend on its organization, capabilities, supply needs, and financial situation. In addition, when a utility evaluates a potential program option, it must examine its technical and economic feasibility, based upon these goals.

An IRP planning process better incorporates the marketplace factors that today's electric utilities must address: competition from other energy suppliers, customers' increasingly complex needs, and cost-effective programs and services.

A Few Examples of Services With Added Value

Table 12-1 lists some service options a utility might offer its customers. These service options meet utility needs but also contain characteristics customers find valuable.

Table 12-1
Examples of Service Options

Utility Initiatives	Product-Related Risk Management	Price-Related Risk Management	Facility Operations Management	Customer Convenience
Back-up Power	Outage Insurance	Fuel Repurchase	Leasing End-Use Equipment	Summary Billing
Contract Bidding	Demand Subscription Services	Contracted Base Rates	Cogeneration Partnerships	Meetings and Seminars
Alternative Energy Investment	Dedicated Service Crew	Bypass Avoidance Rates	Financial Incentives for Efficiency	Equipment Safety Checks
Architectural Assistance	Load Control	Priority Service Rates	Appliance Sales and Repair	Advice on Rate Options
Economic Development Rates	Interruptible/ Curtailable Rates	Future Markets	New Technology Information	Aggregated Chain Accounts
Prepaid Electric Service	Rotating Outage Subscription		Energy Service Company	Personal Account Reps
Spot Gas Credits			Training Customer Employees	Repair Referral Service
5-Year Rate Contracts				
Regional Rate Option				
Energy Service Co. Incentives				
Real-Time Pricing				

The table groups the options in attribute categories, such as risk or convenience. Technology, in particular, plays an increasingly critical role in determining which service options to offer, because so many technology advances add value. Today's service options include equipment that offers enhanced reliability and power quality, more flexible, accurate, and timely metering and information services, and increasingly efficient appliances and productive industrial processes. Utilities must assess each of these technological innovations, determine the constraints that exist, and then decide whether or not to develop the expertise and experience required to offer a service option that includes this new technology. This decision may depend on a number of factors, including customer acceptance and support from trade allies.

For example, several utilities now offer their commercial and industrial customers power conditioning services and products. Power protection is increasingly important to customers as more equipment becomes computer-based. (Computers, cash registers, price scanners, HVAC controls, and other electronic devices can be damaged easily by sudden changes in electric power.) To provide this program, a utility must familiarize itself with the hardware available as well as learn about its customers' specific power conditioning needs. Then, it must develop an effective program that includes services such as power monitoring, diagnosis and analysis of power quality problems, and the installation, verification, and maintenance of power conditioning equipment. Some utilities that currently offer power conditioning programs and services sell the equipment directly to customers and some lease the equipment to their customers.

Many utilities find that new technology improvements help cement their relationships with customers. EPRI is working in this area to develop a unified communication protocol system. This integrated utility communications system allows customers to communicate directly with a utility. For example, Southern California Edison's (SCE) NetComm system consists of three major components: computers located at SCE, a large number of electronic meters located on a customer's property, and a network of packet radios connecting the two. The system allows customers to connect service instantly, question bills, determine costs of energy use, and in general manage their service more effectively. SCE receives important system management benefits as well as develops a better relationship with participating customers.

New technologies are often under-used because of the lack of customer demand (market pull) or the lack of sufficient distribution channel (market push), or both. If electricity consumers want new or improved appliances and ask retailers to provide them, retailers will then ask wholesalers to supply them, and wholesalers in turn will seek manufacturers to produce those products. If consumers fail to act, then the whole string of potential benefits unravels.(12-4)

To create market pull, energy planners must understand how consumers make energy choices. For example, most planners are puzzled to find that customers sometimes shun efficiency even when it is accompanied by attractive economic incentives. In the past, manufacturers and retailers have not considered efficiency to be an important feature in new products, because they have found that consumers rarely decide to make a purchase based on efficiency. The factors that most consistently affect their choices are appearance; safety; comfort, convenience, and control; economy and reliability; high technology features; the need to have the latest equipment; the desire to avoid hassles; and resistance to having utilities control energy use. Because human nature is diverse, the weighing of these factors varies enormously, and retailers must adjust their marketing strategies accordingly. Businesses have analogous concerns, including product quality, production reliability, fuel flexibility, environmental cleanliness, a clean workplace, and low risk.

Conclusions

Today's utilities must serve their customers well. This task is as simple (and as complex) as understanding what customers want and need, and then offering it at prices they will pay.

The preface to this report listed several steps utilities can take to accomplish this task. They bear repeating:

1. *Utilities must begin to recognize that perception—not reality—drives most consumer actions.* Keep this fact in mind when devising programs, services, and their corresponding marketing campaigns.

2. *Utilities must become value-oriented, instead of product-oriented.* "...what the customer buys and considers value is never a product. It is always utility, what a product or service does for him." (Peter Drucker)

3. *Utilities must define and illustrate electricity's value to customers in terms they understand.* They understand it when the lights or the television switch on (or fail to). Utilities should use simple, clear images to convey messages.

References:

12-1   P. Hanser, EPRI Journal, April/May 1991.

12-2   *Impact of Demand-Side Management on Future Electricity Demand: An Update,* EPRI CU-6953, EPRI, Palo Alto, Calif., September 1990.

12-3   F. P. Sioshansi, "Marketing Energy Efficiency to the U.S. Power Industry: Recent Developments and Future Prospects," Presentation at Elmia Energy 92, Jonkoping, Sweden, March 10-13, 1992.

12-4   A. P. Fickett, C. W. Gellings, A. Lovins, "Efficient Use of Electricity," *Scientific American,* September 1990.

Index

Acid rain, 9, 55, 57-59
Adjustable speed drives, 135-136
Aerospace Industries Association, 143
Agriculture, 193-205
  irrigation technology, 198-201
  animal husbandry, 201-203
Alabama Power Company, 15, 194
Amber, ix
American National Standards Institute, 254
American Petroleum Institute (API), 190-191
Amoco Research Center, 50
Apple Computer, 16
Arab oil embargo, 8
Arc furnaces, 24, 35-36, 67, 144, 154
AT&T, 16
Automotive industry, 153-154

Baltimore Gas & Electric Company, 243
Beneficial electrification, 33-35

Bonneville Power Administration, 133
Bossong, Elizabeth, 142-143
Brayton cycle heat pump, 70-71
Brookings Institution, 145

Carnegie Corporation, 145
Carrier Corporation, 20, 60, 62, 133
Cement industry, 43, 138-140
Chemical manufacturing industry, 22, 43-45, 47-48
Chernobyl, 82
Chlorofluorocarbons (CFCs), 9-10, 56-58, 70
Chrysler Corporation, 30, 69, 210
CIE, 50
CLASSIFY™, 242
Clean Air Act, 10, 58
Clurman, Ann, 227-228
Cohen, Stephen, S., 137
Commercial CLASSIFY PLUS, 242, 249-250
Commercial sector, 32-33, 116, 134-135, 237-250

market segmentation in, 237-250
electrification trends, 116
energy efficiency savings, 134-135
Commonwealth of Independent States (CIS), 82
Competition, 10-11, 136, 167-169
Conservation, 35, 59, 108, 112-113, 166, 168, 191, 235, 265, 269
Coors Brewing Company, 70
Council of Economic Advisors, 84
Crude oil cracking technology, 43-44
Customers, electric utility
commercial needs-based market segments, 242-249
demographic and social changes, 223-229
energy needs, 11-12, 15-17, 32-33, 221-255, 258-263, 265, 269-275
residential needs-based market segments, 231-235
role of DSM programs in meeting needs, 263
Customer focused planning, 261

Dairy industry, 27-28, 72-73
Demand-side management (DSM), 13-15, 34, 58, 87, 125-126, 166, 169, 171, 239, 241-243, 250, 253, 259-260, 263-275
current U.S. activities, 265-268, 271-275
definition, 263-264
history of use as a marketing strategy, 264-265
planning in a regulated marketplace, 268-271
utility incentives/cost-recovery methods for implementing, 266-268
Department of Agriculture, 196
Department of Commerce, 84
Department of Defense, 6
Department of Energy, 62, 84, 165
Department of Transportation, 208
Department of Treasury, 84
Dielectric heating, 22-23, 178
Drucker, Peter, 11, 275
Dual-mode electric vehicle (DMEV), 215

Economic changes in U.S., 113-114, 167, 227-229, 237
Economic efficiency in U.S., current, 119
Edison Electric Institute, 224
Edison, Thomas, 7, 95, 99, 265
Electric arc furnaces. *See* Arc furnaces.
Electric infrared (IR) auto coating and drying, 69
Electric heat pump. *See* Heat pump.
Electric motors. *See* Motors.
Electric Power Research Institute (EPRI), 11, 29-30, 34, 60, 62, 64, 125, 132-134, 136, 156, 165, 210-211, 218, 229-234, 242-249, 260-262, 266, 273
Electric transportation, 30-31, 207-218
Electric vehicles (EVs), 10, 30-31, 210-213
Electricity
attributes, general, 19-20
consumption, historic and projected, xi-xii, 34-35, 107, 149, 237-238, 266

consumption, industrial sector, 65, 155-164
economic attributes, 13, 24-25
economic efficiency of, 115-120
environmental impacts, 9-11, 25-26, 30-31, 55-76, 89-91
future role in industrial sector, 164-166
historical background, ix-x, 94-98
impact on industry, 149-171
link with economics, 107, 110, 113-120
link with energy efficiency, 107-114, 124-125
marketing, 7-11, 231-275
relationship to national security, 79-91
relationship to U.S./world economic growth, 116-120
relationship with technology, 3, 33-50, 109
reliability as an energy supply, 252-255
resource use attributes, 25-26
role of DSM, 263-275. *See also,* Demand-Side Management.
social impacts, 93-104, 223-224, 229-231
technical attributes, 12-13, 20-24
total resource efficiencies, 26-33
value, *See,* Value.
value-based programs/services, 14-15
Electrolytic phenomena, 22-23
Electromotive phenomena, 21-24
Electronic cottage, 32, 240
Electrotechnologies, 35-50, 61-76, 142, 144, 169-170, 173-205, 241, 262

Electrothermal phenomena, 21-22
Empire State Electric Energy Research Corporation, 252
Energy conservation. *See* Conservation.
Energy consumption, xi
Energy efficiency, 107-114, 268-269
case study: U.S. versus Germany, 120-124
economics of, 108, 114-120
future improvements in U.S., 124-125, 129
impact of DSM programs on, 125-128
impact on technology, 109, 129-148
national policy goals, 110-111
potential savings, commercial sector, 134-135
potential savings, industrial sector, 135-136
potential savings, residential sector, 132-134
Energy intensity, 109
Energy needs. *See* Customers, electric utility.
Energy productivity, 113-114
Environment, 9-11, 25-26, 55-60, 80-84, 86-91, 170, 176-178, 185-186, 189, 210-218, 240, 263-269
Environmental Protection Agency (EPA), 67

Family units, demographic changes of, 225-226
Federal Energy Administration, 120, 122
Federal Express, 16, 257
Firmographics, 243
Ford, Henry, 153

Freeze concentration, 27-28, 72-73
Fuel switching, 86-91

Gas industry marketing campaigns, 5, 168, 240
Gas Research Institute, 63
General Motors, 30, 68, 210
Glassmaking, 181-186
Gleick, Peter, 80
Global warming, 9, 55-59
Greenhouse effect, 55-59
Gross National Product (GNP), 85, 87-88, 107, 115-118, 130, 136

Hall-Heroult process, 23
Heat pump,
    electric, 5, 29-30, 60-65, 112, 133
    gas, 63-65
    industrial process, 71-73
Hood River Conservation Project, 133
Hoover Company, 95
Huls Company, 44
Hussein, Saddam, 81, 83
Hydrofluorocarbons (HCFCs), 10, 57-58

Industrial sector
    electrification case studies, 173-205
    electrification trends, 116-117
    electricity's impacts on, 6, 164-166, 253-254
    historical perspectives, 149-152
    impacts of regionalization, 152-155
    marketing, 166-171
    overview of recent demand, 155-163
    productivity, 136-146
    research and development, 144-146
    role of electric utility industry, 144-146
Information technologies, 31-33
Integrated resource planning, 268-271
Iron and steel industry
    electricity's impact on, 35-38
    energy requirements of, 38-42
    impacts of regionalization, 154-155
    productivity gains, 138-140, 142-145

Kahane, Adam, 138-140
Korea, 82
Kureka Chemical Industry, Ltd., 44

Least-cost planning, 13, 268-270
Lenin, Vladimir, 152
Load building, 269-270
Load management, 265, 269-270

Magnetic fields, 10
Market segmentation, 231-235, 240-249
    commercial needs-based, 242-250
    residential needs-based, 231-235
Marketing
    definition, 260
    electric utility efforts, 7-9, 15-18, 250-251, 257-275

relationship between service and value, 257-258
role of DSM in utility efforts, 264-265
to industry, 166-171
under regulation, 259-260, 268
Marketing mix, 250
Massachusetts Institute of Technology (MIT) Commission on Industrial Productivity, 136
Materials production, 66-69, 156-164
Materials fabrication, 69-70, 156-164
Maximum technical potential (MTP), 129
Medical electronics industry, 45-46
Minimills, 35, 37-40, 144, 154-155
Mississippi Power, 15
Montreal Protocol, 10, 58
Motors, 21, 65, 135-136, 151, 192

National Academy of Sciences, 90
National Appliance Energy Conservation Act of 1987, 61
National Energy Policy Act of 1992, 58
National Personal Transportation Study, 208
National Science Foundation, 89
Needs, customer energy. *See* Customers, electric utility.
Nonferrous metals industries, 42-43
Nordstrom, 16
North American Electric Reliability Council, 125, 165
Nuclear Regulatory Commission, 89

Oak Ridge National Laboratory, 76

Office of Management and Budget, U.S., 84
Oil. *See* Petroleum refining.
Ontario Hydro, 253

Pacific Gas and Electric Company (PG&E), 252-253
Pacific Power & Light, 133
Persian Gulf War, 79-81, 83-84, 91
Personal rapid transit (PMT), 215
Peters, Tom, 16
Petroleum refining, 186-193
  electric powered refinery technologies, 192-193
Pinch technology, 73
Plasma arc technology, 36-37, 43-44, 48
Power quality, 254-255
Process industries, 27-28, 70-73, 75, 156-164
Productivity, 113-114, 136-146
Public Utilities Regulatory Policies Act of 1978 (PURPA), 168
Puget Sound Power & Light/Puget Energy Services, 268
Pulp and paper industry, 6, 138-140, 173-179
  case study of 138-140
  chemical pulping, 173, 176-178
  mechanical pulping, 173-176
  paper drying, 178-179
Pulse combustion furnace, 62-65

Reagan Administration, 141
Rebate programs, 263-264, 270
Reconstruction Finance Corporation, 195
Red River Valley Electric, 15
Residential sector,

electrification trends, 116
energy efficiency savings, 132-134
energy needs, 229-235
Resource Conservation and Recovery Act, 67
Roosevelt, Franklin, 194
Ross, Charles, xi
Rural Electrification Administration (REA), 194-195

SNIA Fibre, 50
Solvay, 50
South Dakota State University, 202
Southern California Edison (SCE), 273
Soviet Union (former), 82
Static electric fields, 10
Steel industry. *See* Iron and steel industry.
Steel Survival Strategies Forum, 143
Sugar refining, 6

Target marketing, 231-232
Taunton Municipal Lighting, 15
Technology, relationship with value, 273-274
Telecommuting, 32
Tennessee Valley Authority, 195-196, 253
Textile industry, 152-153
Thermal energy, 23
Thermoplastic composites, 6
3M Company, 71
Toffler, Alvin, 227-228
Transportation. *See* Electric transportation.

Ultraviolet radiation (UV) for curing can coatings, 70
Unbundled services, 251
Union Carbide Corporation, 43-44
United Nations, 81
United Parcel Service, 16
USX Corporation, 142
Utilities, introduction of into U.S. life, 94-95, 149-150

Value, general concept of and relationship to electricity, xii-xiii, 3-6, 9-17, 19, 221, 251-255, 257-274
Vietnam, 82
Volatile organic compounds (VOCs), 69-71
Waste and water treatment, 66, 68, 70, 73-76
Wisconsin Electric Power Company, 15, 243
Women
  demographic changes, 225-226
  electricity's impacts on, 98-104

Yankelovich, Skelly and White, 227

Zysman, John, 137